Climate Strike

Climate Strike
THE PRACTICAL POLITICS OF THE CLIMATE CRISIS

DEREK WALL

MERLIN PRESS

The Merlin Press Ltd
Central Books Building
Freshwater Road
Dagenham
RM8 1RX

www.merlinpress.co.uk

First published 2020

© Derek Wall, 2020

Cover photo © Sean Hawkey, 2019

All rights reserved. No part of this publication may be reproduced, stored in a retrieval system, or transmitted in any form or by any means, electronic, mechanical, photocopying, recording or otherwise, without the prior permission of the publisher.

The author's moral rights have been asserted.

ISBN. 978-0-85036-764-5

A CIP record of this book is available from the British Library

Printed in the UK by Imprint Digital, Exeter

Contents

Introductory statement from Sean Hawkey	vii
Acknowledgements	viii
Chapter One: From Icy Puddles to a Burning Planet	1
Chapter Two: Hothouse Earth	19
Chapter Three: Another Green World	33
Chapter Four: Green Parties and Climate Change	51
Chapter Five: Trade Unions for Life on Earth	64
Chapter Six: Extinction Rebellion	75
Chapter Seven: The Climate Accelerationists	100
Chapter Eight: The Unconscious, the Imaginary and the Real	123
Chapter Nine: Don't Trust the State	139
Chapter Ten: Green Lenin, Green Machiavelli	151
Further Reading	174
Abbreviations	176
Glossary	178
Bibliography	180
Index	202

'We are at the midnight hour, and it's eco-socialism or death.'
Kali Akuno (2019)

Dedication
In memory of Carole Chant and Ted Knight

Introductory statement from Sean Hawkey

I feel it's painful to see where the world is going. These last ten years, we've been witnesses to climate change advancing unchecked, the destruction of the environment we depend on, and if change is coming, it's very slow. We've seen the theft of people's lands and resources at any cost, so much it has become normalised already.

And the power of truth in journalism, politics and law is waning. I think it's a dangerous moment for democracy, justice and it makes dealing with climate change a lot harder.

Sean Hawkey

Sean Hawkey is a photographer and kindly provided the image for the front cover of this book. You can find his work here https://www.hawkey.co.uk

Acknowledgements

My wife Emily Blyth, yet again, read my drafts carefully. Richard Martin also read the entire draft and made numerous useful suggestions. Both Emily and Richard deserve thanks for all their intelligent care for this book, it could not have been written without them. Ian Angus, Oscar Berglund, Chris Jarvis, Darrell Hannah, Marc Hudson, Heidi Ravven and Alan Thornett also made useful suggestions for editing and improvement. Sean Hawkey was kind enough to let me use his photograph for the front cover. Andrew Newman's recommendation of Behringer's *A Cultural History of Climate* provoked much thought. Anthony Zurbrugg has been a patient and supportive editor at Merlin Press. Finally, I would like to thank Adrian Howe for his work in producing this book.

Chapter One

From Icy Puddles to a Burning Planet

'Well, you should write a novel for the workers,' Says Lenin to Bogdanov, 'on how predatory capitalists have despoiled the Earth and wasted all its oil, iron, lumber and coal. That would be a very useful book.' (Wark 2015: 8)

Once upon a time I saw icy puddles. Attending primary school in Wiltshire in the 1970s, I would be charmed by the ice patches often to be found in Autumn and Winter, shining in the sun in the playground of Frogwell Primary in Chippenham, Wiltshire and later at Corsham School. Memory deceives but nearly fifty years later and just one county away, when I get up on a winter morning in Berkshire, ice is rare. My perhaps trivial recollection is a faint personal reflection on a warming world. As I write, carbon dioxide levels are climbing ceaselessly, temperatures are rising, and extreme weather is increasingly the new normal. Since 1980, as a politically precocious 14-year-old, I have been committed to green politics. Worried about the state of the world, I have written, taken direct action, read until my eyes were raw and strategised on a daily basis. All the effort has seemed both prophetic but also perhaps quixotic; climate change and other ecological ills are taking hold not of the future but the present. The aim of this book is to advance strategic debate.

A central claim that I make, and I am far from original in doing so, is that climate change is a product of the entire social and economic system within which we exist: in a word, capitalism. Challenging capitalism is essential to challenging climate change. This leads to a number of very difficult issues. How can we create an efficient non-capitalist economic system? How can we promote political action to end capitalism and to do so quickly? Given what appears to be the rather abstract nature of climate change, how can individuals be mobilised to act? The future of humanity, for some, apparently generates little or no concern? What are actions that can be taken now within the framework of capitalism to at least slow rising emissions?

The challenge is that if the economic system, indeed the basic social system, needs to be transformed to protect humanity and the rest of nature, this cannot be achieved easily or quickly. Climate change and less visible environmental threats demand almost instant action, so interventions are necessary immediately. Therefore, the politics of climate change needs to take a seemingly contradictory approach, intervening directly and immediately to slow emissions but working in a more fundamental and long-term way to promote the creation of a different way of life with all the complexities of institutions and practices that this demands. Systems cannot be replaced with planned utopias but instead interventions can begin to move such systems in a different direction.

There are a number of responses to climate change that appear problematic. One approach is to do little or nothing. Despite the international meetings and global frameworks put in place to reduce greenhouse gas emissions, this is perhaps the dominant one. Emissions rise, action seems tokenistic and inaction is common; typically in the UK, climate change has little or no policy priority. Even though coal fired power stations are phased out, other aspects of policy such as road building seem to show no awareness of the challenge of climate change. While more roads are built in the UK, rail fares rise and cycle path provision here is derisory.

A second approach, fast growing as I write, is to work to make the problem worse. Right-wing leaders like Donald Trump and the Brazilian President Jair Bolsonaro have labelled climate change a hoax, doing all they can to rip up environmental regulation. Trump declared his love of coal, while Bolsonaro has advocated clear cutting the Amazon. At the 2018 COP Climate Change Conference in Poland, a coalition of the US, Kuwait, Russia and Saudi Arabia vetoed the acceptance of the scientific report prepared for the event (McGrath 2018a). At future events they will be joined by Brazil. The rise in governments actively working to accelerate climate change is alarming. Incidentally, COP is an abbreviation of Conference of the Parties, the supreme decision-making body of a United Nations convention.

Another perspective born out of the need to resist inaction and acceleration is to call for strong action from governments and the international community. While this is laudable, we need to be conscious that broadly drawn demands for action may not translate into effective solutions. A major claim of this book is that calling for action is insufficient; climate change and other ecological ills are complex problems, requiring radical social and economic change. The road to disaster is paved with unintended consequences; policies to deal with climate change, for example the international carbon market, can be criticised in a number of ways (Lohmann 2006).

In turn, I am sceptical that there is 'one solution'. For example, a more vegan diet may be laudable, but climate change action cannot be reduced to one measure, whether dietary or not. The critique, in turn, of individual action will be made. Structural transformation is vital, reducing personal emissions is less important than creating good public transport systems that allow us to have less impact. Thus political change is necessary.

Politics relies on stories of heroes and villains, right and wrong and the construction of simple solutions. However political change is complex and revolutionary transformation is unlikely to be easy or neat. The stories which are told can oversimplify the tasks that are necessary, lead to false solutions and translate problems into familiar but misleading narratives. In contrast, however, a conceptual understanding, based upon rational analysis, can lead to a politics which is too abstract to be effective in rousing us to action. This tension between narrative and analysis demands examination. In turn, it can help us to understand the basis of movements like that of Trump that are otherwise both threatening and difficult to comprehend. I will attempt to outline my thoughts in the following pages. Yet any one individual's thoughts are of course, largely, common property, the product of social learning and collective endeavours. They are perhaps never really personal. My approach has been strongly influenced by the work of two individuals, Elinor Ostrom, the first woman to win a Nobel in economics, has produced a powerful body of work on ecological economics, climate change and the commons (Wall 2017). The second is Hugo Blanco, an indigenous leader and ecosocialist revolutionary. His thinking has shaped my approach to militant ecological politics (Wall 2018). In turn, a number of thinkers, focused on literary theory and philosophy, with a shared materialist outlook, most particularly Étienne Balibar, William Lewis, Warren Montag, Heidi Ravven, Jason Read and Ted Stolze, have deepened my understanding of practical political action. They are, of course, not to blame for my misunderstandings and mistakes.

Radical change is necessary but difficult. Often all paths to political change look blocked, which is why some deep thinking is needed on the question of strategy. I am not here to provide a short rousing slogan that encapsulates what is to be done. However while there are no guarantees, shifting green politics from posing demands for an ecologically sustainable future to the task of asking how we get to an ecologically sustainable future is necessary. My polemical demand is the demand to think, to ask, to explore and to act as effectively as possible. Green politics poses such important questions, of tackling potentially devastating ecological threats, including climate change, that green political strategy becomes a matter of extreme importance. This is perhaps my most passionately and deeply held conviction.

Thinking about political action with Jean Cavaillès

A French mathematician who was killed in 1944 might appear an unlikely source of inspiration for climate change action. A better start might be to look at the words and work of Roger Hallam, the anarchist student of social movements, who helped in part to kick start the Extinction Rebellion movement (Hallam 2019). Or why not Yeb Sano, the climate conference delegate who broke down in tears at the thought of what extreme weather was doing to his country, the Philippines. Indeed Greta Thunberg, a Swedish 16-year-old, who attacked politicians as too immature to tackle climate change at the 2018 COP conference, might provide a better example than Jean Cavaillès. Hundreds of more worthy figures might be conceived than Jean Cavaillès, however I feel his short life provides both inspiration and instruction.

Jean Cavaillès, born in 1903, was a philosopher of mathematics. His most important publication *Sur la Logique et la Théorie de la Science* (On Logic and the Theory of Science) was written while he waited in prison to be executed. He combined highly abstract work on the philosophy of mathematics and the implications of this for human reasoning, with a short-lived life as one of the most militant and practical opponents of the Nazi occupiers of France in the Second World War (Canguilhem 1996). He combined precise action against an oppressor with deep and abstract thought. In both regards he is significant. While context is all and fighting the Nazis is not the same as fighting climate change, I think he has something to teach us.

During the early years of the twentieth century the study of mathematics was transformed and perhaps threatened by the work of Georg Cantor's set theory. Cavaillès noted that the implications of this form of mathematics were so radically counter-intuitive that they showed the extremely limited nature of human consciousness. What we can understand based on our observation, our perhaps common sense, was misleading. To repeat a well-known phrase, it might be argued that reality is not only stranger than we think but stranger than we can think. Cavaillès noted that what was necessary at the boundaries of mathematics and the philosophy of science was not an engagement with 'consciousness' but 'concepts'. Observation gets us so far, but rational analysis based on formal logic, gets beyond what we can consciously see, hear or feel (Peden 2014). I am not suggesting that all aspects of the politics of climate change are so abstract and challenging that we replace common sense empirical observation with a rationality based on formal abstraction. However I strongly suspect that some aspects of how we bring about change in human society so as to combat environmental problems are likely to be difficult to grasp, beyond our intuition, and demand a break with strongly held assumptions. The very translation of concepts into images that we can easily grasp distorts their nature. Part of our work must be conceptual, understanding the science of both society and the rest of nature, or at least trying to use our rationality to become a little less ignorant.

Cavaillès' mixture of rationality and militancy provides an example worthy of close study. Where things are so strange that we might fail to acknowledge their existence, he inspires us to understand them more profoundly. Abstraction was one side of his work; engagement of the most personally costly kind was another. Writing about his legacy, the historian of philosophy Knox Peden noted,

> In addition to being one of the leading philosophical minds of his generation, Cavaillès was one of the most active leaders of the French Resistance during the Second World War. (Peden 2014: 18)

Joining the French army as an officer in 1939, Cavaillès was captured in 1940. He escaped prison and resumed his post as a professor of philosophy at the University of Strasbourg. He quickly established the *Libération-Sud* to resist the Nazi occupation. In 1941 he was appointed as a professor at the Sorbonne, combining secret missions across France, with his academic work. Arrested in September 1941, he escaped once again and made his way to London, where he called upon the Gaullist and Communist wings of the Resistance to focus on strategy. While none of us can reject ideology, we are all caught in webs of belief and commitment; Cavaillès sought to stress, perhaps foolishly, a basis of rationality for resistance. A close follower of the seventeenth-century Dutch-Jewish philosopher Spinoza, incidentally, often lauded by environmentalists for his arguments focusing upon 'nature', Cavaillès argued that it was a matter of logic to oppose the Nazis. He noted 'I'm a Spinozist, we must resist, fight, and confront death. Truth and Reason demand it.' (Peden 2014: 19)

While armed struggle does not obviously spring to mind in combatting climate change, and the political value of the work of Spinoza is subject to intense debate, Cavaillès' rousing call that 'we must resist, fight and confront death' is a starting point for challenging climate change. Climate change threatens life and we must resist. Returning to France, he entered the German submarine base at Lorient in Brittany, and placed and detonated explosives. Successful in a number of similar missions, he was eventually re-captured in a Paris street in September 1943. Imprisoned, he wrote on philosophy, science and mathematics. He was executed by firing squad on 17 February 1944. Peden notes that the philosopher and sociologist Raymond Aron recalled Cavaillès telling him:

> I am a Spinozist; I believe we submit to the necessary everywhere. The sequences of the mathematicians are necessary; even the [historical] stages of mathematical science are necessary. This struggle we carry out is necessary as well. (Peden 2014: 19)

I am not sure I agree with Cavaillès in such a view of necessity, but in fighting climate change, urgent and passionate commitment needs to be guided, where possible, by rational analysis rather than anger or despair. Cavaillès' analysis and life inspired both passion and commitment. His friend and fellow resistance fighter Georges Canguilhem, writing in *The Life and Death of Jean Cavaillès*, stated 'A philosopher-mathematician loaded with explosives, lucid and reckless, resolute without optimism. If that's not a hero, what is a hero?' (Canguilhem 1996: 35)

Nature and ecology

I am not aware that during his short and busy life Jean Cavaillès discussed environmental matters, but as a Spinozist he must have been aware of the philosopher's thoughts on nature. Spinoza noted in a pithy but radical statement that human beings have no kingdom within nature (Spinoza 1930: 104). By this he meant that we were part of nature and subject to the rules of nature. Human culture and society for Spinoza is a product of nature. How we understand our relationship to the rest of nature is important in understanding our response to climate change. It is, like so much else, far from simple or self-evident.

Attempts have been made to build a philosophy of deep ecology on the basis of Spinoza's statement, for example, by the Norwegian philosopher and mountain climber Arne Naess (1993). This is perhaps problematic because Spinoza, in his words, felt that everything was 'God or Nature', so if everything from a carpet to a cosmos is nature, this gives us nothing to distinguish. If everything we do, however destructive, is part of nature, being part of nature does not provide an ecological ethos.

Spinoza points us usefully, I feel, in another direction. We think of ourselves as autonomous, conscious and largely independent to make decisions. Yet according to Spinoza, every decision we make is shaped by the forces of nature; our independence is illusory. This sounds depressing and may be fatalistic, but it has a positive implication perhaps in that the more we understand of the forces that shape us, the more we can influence them (Ravven 2013). The politics of climate change must not rest on an assumption that we are fully in control and fully rational. The forces that shape every one of us are not always immediately visible, the populist right have been good at building support by drawing upon narratives and emotions. Understanding the pull of such forces is a part of the process of resisting them.

Arguments from nature are also problematic in that they have often tended to be both foundational and conservative. Foundational, in that like the words 'God' or 'science', nature acts as a warrant or guarantee. If something is natural, it is correct, and the word 'nature' is used to close down debate. Conservative, because not only is biology seen as destiny, but it is so often used to suggest that culture must be based on assumptions of inevitability. Homosexuality has been condemned by bigots as 'unnatural', diverse sexual identities have been challenged as wrong because of the apparent binary nature of men and women. Yet a little research suggests that the assumption of a foundational and conservative nature is false, nature is astonishingly diverse and varied. Scientific discussion of sexual identity, for example, suggests that biology is more flexible than popular discourse might suggest (Sun 2019).

The use of nature as a basis for green politics has been fiercely challenged. The author and academic Tim Morton even entitled one of his books *Ecology without Nature* (Morton 2007). What we think of as nature, in Britain the fields and hedgerows, or in the USA the 'wilderness' of the great National Parks like Yellowstone, is a particular product of human interaction with other parts of nature. Hedges are a part of a class society, used to enclose land in feudal times, and while we should not root them out, other forms of ecological productivity and beauty can be constructed. The Amazon and other rainforests have been inhabited by humans and shaped by human activity for thousands of years (Rostain 2012: 22). Our conscious observation and valuing of a beautiful landscape needs to be more critical, based on an awareness of the concepts behind the trees, both biological and social, that have created them.

I am not, in questioning a romantic commitment to particular forms of a natural environment, advocating instead an ecological modernism based on technological revolution.

The domination by one species, based on limited knowledge, is likely to give rise to dramatic unforeseen circumstances. However, in valuing nature, we need to be aware of its contingent and changing nature. Even without the human part, nature has changed dramatically as a result of geological and sometimes even extra-planetary forces.

I agree with Morton that 'ecology' is perhaps a more useful focus than 'nature'. Spinoza's striking and often prophetic work seems to suggest a philosophy of ecology. The term ecology, coined as recently as 1866 by the German biologist Haeckel, is the science of relationships (Stauffer 1957: 140-1). It charts the interaction between different species including humanity, illustrating the diversity and productivity of life. So while I will often use the term 'nature', I am conscious that an ecological and relational understanding is important.

Our understanding of ecology both within specific ecosystems and more conceptually has been transformed since the nineteenth century, a narrative of different environments moving through stages (*seres*) towards a stable equilibrium, has been challenged by a more chaotic understanding (Kingsland 2005).

Ecological matters are often ignored in politics and, although there are exceptions from John Stuart Mill's critique of growth (Mill 1888), to Engels' concern with water and air pollution (Engels 1958), the green element is often absent from historical political discourse including socialism and liberalism. However, rather than an engagement with difficult science, a response has often been to substitute an ignorance of ecology with a rather sloganist appeal to protect the environment. Promoting the good and the green, while necessary, is not enough. Ecological politics has to be based on a sophisticated understanding of natural sciences such as biology, hydrology, geology, etc. It is also necessary to pursue a sophisticated understanding of social science if human beings are to agree to act and build the institutions and practices that produce effective action. This was one of the insights of that great political economist

Elinor Ostrom (Ostrom 1990); she argued that in dealing with climate change we need to combine natural and social sciences to understand both human behaviour and ecological factors.

The political economy of climate ecology

Although climate change is complex, the basic mechanism driving rising temperatures is, in essence, both simple and long understood. While John Tyndall is associated with the discovery of the greenhouse effect concept, an American woman Eunice Foote is now recognised as making the link between carbon emissions and climate change in 1856, with a presentation to the American Association for the Advancement of Science in Albany, New York (Jackson 2018: 158). Carbon dioxide and other Earth warming gases, such as methane, help to keep heat from escaping from the atmosphere. Without this greenhouse effect, the Earth would be a cold sphere of rock. Huge quantities of CO_2 laid down in deposits millions of years ago have been released, as coal, oil and gas have been extracted and burnt since the industrial revolution of the nineteenth century. If this continues, it is likely that feedback mechanisms will accelerate the heating effect. For example, rising temperatures thaw the permafrost in Siberia and Canada. These release methane, and methane is a more potent climate change gas than carbon dioxide, having twenty-five times the effect of 'CO_2 over a 100-year time horizon' (Reay, Smith and van Amstel 2010: 2). Happily one recent report that tracked methane from a previous period of warming after an ice age, suggests that this may be less of a problem than once thought. Previous thaws in permafrost apparently led to little methane entering the atmosphere. One reason may be organisms in the soil a few centimetres on top of the permafrost that feed off the methane and break it down. A researcher noted, 'We can probably thank those little bacteria for saving our bacon' (Monroe 2020). This not a cause for complacency, some methane and more CO_2 is still likely to be emitted.

Climate change is a visible and much debated environmental danger but there are numerous other threats to our environment from the loss of species, to ocean acidification, to the spread of plastic waste and nuclear waste (Vollmann 2019a). The causation is human action, but human action is shaped by cultural, economic, political and social forces. The causation is ultimately, but not exclusively, a product of capitalism. A capitalist economic system, it has been well established by both those who criticise and celebrate it, tends to promote economic expansion. Economic expansion has tended to be based on increased consumption, production and waste. While capitalism has been associated with technological innovation, which, in turn, may lead to more efficient use of resources, it is also associated with increasing extraction and the simplification of biological systems. To revisit an argument often made, capitalism is a system that demands we consume more, while ecological systems are best sustained when we consume less (Kovel 2007). Capitalism may also encourage the production of what is noxious, for example, weapons of war, rather than what is healthy for both humanity and the rest of nature.

Climate change is not based on individual choice; our choices are not entirely autonomous but shaped by the society we live in. In a capitalist society, economics dominates, and this is the economics of more not the economics of enough. So while we might eat less meat, or consume more consciously in other respects, the systems is organised to keep us on the treadmill of consuming at an increasing rate. The economics of capitalist growth is over determined; it is part of a web of interlocking causes that can also be described as effects. One is cultural: we are used to seeing humans as independent and superior, we have a mindset, at present, perhaps, that promotes domination of the rest of nature. Another is our current dominant form of property rights. While the individualistic principle of personal ownership has perhaps been submerged by corporate control and all manner of opaque instruments, property is seen as serving personal need in the short term. The assumption that we should leave what is owned in a good state for future generations is absent from current principles of ownership.

It is often enlightening to examine the contradictions in a text; it can be equally useful to examine the silences. There can be a silence when it comes to the reality that climate change is a product of imperialism. An instructive text is Frantz Fanon's *The Wretched of the Earth* (Fanon 1983). Colonial imperialism was an exercise in cruelty and economic expansion, based on the exploitation of labour; it also exploited of natural resources. The Belgian assault on the Congo is one example, a population enslaved to extract rubber. The Indigenous are on the frontline of the destruction of the environment to extract fossil fuels. In countries like Peru and Ecuador the drive to extract oil and natural gas involves the continued displacement of Indigenous people in the twenty-first century.

The history of fossil fuel extraction is closely linked to a history of imperialism. Imperialism is an economic concept developed by thinkers such as John Hobson, Rosa Luxemburg and Lenin, suggesting that European countries were innately expansionary. To avoid economic crisis, colonisation allowed the capture of new markets for goods, cheap labour and new resources (Brewer 1980). The Swedish ecosocialist author Andreas Malm, in his book *Fossil Capital*, notes that hunger for coal and other resources helped drive the expansion of the British Empire (Malm 2016). One example of this he describes is the exploitation of coal on Labuan, an island just above the tip of northern Borneo, where an area of tropical forest was cut to make it possible to dig, transport and burn the fossil fuel resource. The coal was needed for imperial steamships, transporting troops, missionaries and merchants. Malm describes it as,

> a distinctly British invention, most simply defined as an economy of self-sustaining growth predicated on the growing consumption of fossil fuels and therefore generation a sustained growth in CO_2 emissions. The coal of Labuan had never before been connected to any such pursuits. The native population knew about it but had left most of it untouched:

only with the arrival of the British was the coal hauled into a circuit that expanded by setting it on fire. (Malm 2018: 20)

Yusoff argues in *A Billion Black Anthropocenes Or None* that damage to the climate rests on centuries of racist exploitation during colonialism (Yusoff 2018). Racism, as an aspect of imperialism, was closely connected to the expansion of financial systems that both funded and grew with such exploitation.

Capitalism has continued to evolve since the end of the British Empire; however the ghost perhaps never quite dies. For example, British banks continue to be a major source of finance for the expansion of new coal mines (Duncan 2019). Capitalism, in its different varieties, depends on growth in production and consumption. It has a tendency to degrade nature and use more fossil fuels. Climate change demands an end to capitalism. An alternative is a system of commons, where resources are owned by communities and ownership of products is increasingly communal too. Goods can be made to last longer; systems of social sharing can be introduced to promote prosperity without economic growth. Exploring the economics of enough is important but this is a question not just of economics but of cultural, social and political forces. Our capitalist economic system is deeply embedded, so much so that its assumptions are lived by all of us, and imagining a different kind of economy is difficult. Changing a system is a matter of revolutionary transformation, along with a host of changes in practices and institutions. In turn, while the critique of the market is strong, the critique of a planned, mechanical state system also has validity.

Many appeals for action on climate change treat human beings as a homogenous group. Our diversity and differences are ignored as we are all asked to change our lifestyles, consuming less energy and doing our bit. Yet we are not all equally responsible for climate change. Human society is divided by gender, ethnicity, class and other factors, with some groups having more environmental impact than others. Those of who are most privileged may do the most damage. The 0.1 percent who travel by Learjet generate more emissions than the poorest who have no access to air travel. A Learjet trip from,

> Aspen, Colorado to San Francisco – 1,386.6km (861 miles) – would, according to our calculations, have CO_2 emissions of 4,411.8kg. By contrast, driving the average car 10,000km (6,213 miles) over a year would emit about 1,600kg of CO_2.
>
> But to live an affluent lifestyle, one does more than hop on a jet. There are the houses, cars, boats, clothes, jewellery, technology – money is no object so there is no limit to the amount of objects one can possess. (Wells and Touboulic 2017)

The very rich are in the best position to protect themselves from the effects of climate change. They can afford to relocate if their beach homes are threatened by rising tides or their golf courses need more water to remain lush and green during drought. The rich have air conditioning and will hardly notice if food prices rise. Indeed the political theorist Jodi Dean has argued that the super-rich are preparing already and can quite happily exist with conditions that might be seen as otherwise catastrophic:

> The super-rich are grabbing more and more of the social product. They are buying up more land, hoarding. Housing is built so that the rich have more, and the poor have nothing. And the thing is, the rich, the Right, have people, forces, throughout the government and in the streets. They fight tenaciously and refuse compromise. (Dean personal comment)

Thus another dimension of the understanding that capitalism drives ecological destruction is that such destruction accelerates different inequalities including those based on ethnicity, gender and social class.

As we have seen there are powerful forces emerging that are working to make emissions higher and to create, in the longer term, some new and even more destructive system. In recent years, the far right have apparently been winning more victories and mobilising more people on the basis of vivid images and effective slogans. The victories of politicians like Trump and the Brazilian president Bolsonaro cannot be dismissed as based on manipulation alone. Their demands talk to the desires of many of us. Discussion of effective mobilisation by political parties and social movements is necessary; without attending to this, those advocating action on climate change, even of a modest and reformist nature, will fail.

Ecologies of action

Best known for giving the world the iconic detective Sherlock Holmes, Sir Arthur Conan Doyle also invented a character who intimately represents everything wrong with humanity's relationship with the rest of creation. Professor Challenger is a scientist found in a number of Conan Doyle's stories, including *When the Earth Screamed*. Beneath a veneer of rationality, he is inspired by cruelty and the desire to dominate others. Professor Challenger is convinced that the planet Earth is a living organism and we humans are like algae covering its skin. To prove his point he drills a shaft deep into the Sussex countryside and shoots a huge metal dart into the interior of the planet. A horrible cry of pain is emitted along with a gush of a dark oily substance (Pindar and Sutton 2014: 1). This is a fiction which, conceivably, tells the truth about our separation and arrogance from the rest of nature.

The French psychoanalyst, philosopher and Green Party member Félix Guattari, who incidentally was roused by reading *When the Earth Screamed*, argued that in challenging ecological destruction, we should consider the importance of three ecologies. Environments, organisations and individuals are all made up of networks that interact, they are thus ecologies (Guattari 2014). In a sense, what I am trying to promote in this book is an understanding of ecologies of action. Change and indeed conservation are about the operations of networks, systems if you like, that can be understood ecologically. The individual is only ever a part of a wider system, and political action cannot be understood as simply taking power and commanding things to be different. The politics of ecology demands changing a system, capitalism, into a new system. Change in this form is based on interactions, there is not one simple way of solving a discrete problem, instead systematic transformation is the necessary task.

In thinking critically about how we effectively challenge climate change I am keen to avoid erecting a dogmatic system, defending a fixed doctrine or calling out others. Too often, criticism on the left can be destructive of movements on the left (Fisher 2013). Concepts can be used to aid productive change, so ideas matter, they have a material force. However, theory requires debate and respect for difference is part of the process.

Book Outline

Chapter Two challenges denial, discusses the basic science behind climate change and outlines the negative effects of a warming world. Chapter Three suggests that to deal with the climate crisis we must ultimately construct a post-capitalist economy that promotes prosperity without destruction. Chapter Four discusses the role of Green Parties in political change, with an assessment of their strengths and weaknesses (especially around strategy and tactics). Chapter Five looks at the contribution trade unions can make to tackling the climate crisis. Chapter Six puts social movement mobilisations such as Extinction Rebellion under scrutiny. Chapter Seven looks at the dangers of far-right mobilisations gaining votes through racism and exclusion, in turn introducing policies to accelerate climate change. Chapter Eight examines the politics of persuasion, showing how challenging effective climate change communication can be. Chapter Nine argues that we can't trust the state; policies to tackle climate change need to be both effective and socially just. States may deliver both injustice and incompetence, so their actions need considering with some caution. Chapter Ten concludes by suggesting immediate action we can take now together with strategies for more fundamental transformation of society. Throughout this book I am trying to promote Cavaillès' understanding that effective action rests on an examination of relevant concepts. The deeper we think about what we do; the more we reject the visible and the obvious, and examine things more critically, the more effective we are likely to be.

Chapter Two

Hothouse Earth

> *The Cooling* will be controversial, because among scientists most of the matters it deals with are hotly debated. There is no agreement on whether the earth *is* cooling. There is not unanimous agreement on whether it *has* cooled, or one hemisphere has cooled and the other one warmed. One would think that there might be a consensus about what data there is – but there is not. [...] One extreme view envisions a 'snow blitz' beginning of the ice-age climate, only a few years long, and a rapid growth of continental glaciers. (Bryson 1976: xi)

Despite evidence of higher temperatures, rising CO_2 and long-established science linking emissions to climate change, denial and scepticism are common. Some still argue that the science is far from settled. This chapter considers challenges to the science, outlines the arguments for linking rising temperatures to the greenhouse effect and discusses the dangers that climate change poses to humanity and the rest of the nature. Science isn't the only, or even main, reason for denial, but examining the science is important.

Cooling on warming?

The complexity of climate history might suggest rational reasons for doubting the mainstream. As an undergraduate studying environmental archaeology in the 1980s, I remember being faced with a daunting reading list on the topic of the causes of the ice age(s). The last period of glaciation which peaked approximately eighteen thousand years ago, turned large parts of the world into icy wasteland (Rapp 2012:1). One rhetorical tactic of sceptics has been to point out that during the 1970s, far from being concerned with global warming, some scientists were fretting about the possibility of planetary cooling, a return to a full ice age. Mike Hulme, author of *Why We Disagree about Climate Change*, and a founding director of the Tyndall Centre for Climate Change Research, noted as an undergraduate geography student at the University of Durham: 'I well remember being taught at that time about claims of an approaching and seemingly imminent Ice Age, which struck me then, as they still do today, as being far-fetched.' (Hulme 2009: xxix) The seemingly mythical book *The Cooling* (Ponte 1976) can be a little difficult to find in bookshops but I managed to track down a copy. A close reading of this text is instructive. The language of concern over cooling is not so different to that of the language of anxiety in more recent decades over warming. The danger to humanity, it was argued, from cooling, is not a future threat from glaciers marching half-way towards the equator, instead it is immediate and can be measured:

> in lost bushels of wheat and corn and in barrels of oil diverted from other purposes to heat homes and factories and offices. In the time it takes you to read this book, at least a thousand people will have starved to death *because of* the impact climatic instability already has on food production. But by some indications the cooling has scarcely begun. (Ponte 1976: 90)

Lowell Ponte is an American journalist and political commentator, now firmly on the right, but his political stance in *The Cooling* is less clear. Interestingly even a casual reading of his book provides evidence that there was no consensus on cooling in the 1970s, despite the claims of some climate sceptics in recent years. The preface by the climatologist Professor Reid Bryson indicates that scientific opinion was divided as to whether cooling had occurred and, if it had, what was driving it.

Despite writing a popular science book on the topic, Ponte seems somewhat conflicted too. Far from cooling being the consensus in the 1970s, as sceptics suggest, even those who toyed with idea like Ponte, seem to have been ambivalent. However, his broad argument was as follows. First, there has been evidence of temperatures dropping, frosts in Brazil disrupting coffee supply and such like. In turn, the occurrence of ice ages, divided by warmer periods, suggested that the glaciers might at some point make a return. A combination of global dimming, caused by increased particles from burning fossil fuels, and the influence of the Milankovitch cycles (discussed later in this chapter), was likely to lead to falling temperatures. This, in turn, would lead to feedback mechanisms that would cool the Earth further.

Feedback mechanisms are a major and important topic of climate change science. Ponte argued, and in doing so follows a well-established logic, that falling temperatures would lead to more snow and ice cover. This would make areas of the planet whiter and lead to more light being reflected into space, this would feedback into further falls in temperature, leading to more snow cover, etc. This albedo* effect is indeed well known (see glossary).

Ponte argued that such cooling would have a devasting effect on human society. Taking up a Malthusian theme, common in the 1970s, he argued that over population was already putting human food supply at risk. Cooling would reduce crop yields and starvation would increase. Ponte is clear that while the jury was still out on cooling, it could lead to appalling consequences and action was necessary. The action needed was less clear; Ponte didn't advocate burning more fossil fuels, because he felt that by adding more dust to the atmosphere this might increase cooling via global dimming. Ponte speculates as to the possibility of major technological intervention to manipulate the weather, an example of what is now known as 'geoengineering'. With the Cold War between the Soviet Union and the USA in full flow in the 1970s, he suggested that either side might be tempted to use

---

* A glossary at the end of this book defines Albedo and other terms.

climate weapons of some kind.

Ponte's book is interesting but poorly referenced. With only a select bibliography at the end of the book, it involves a fair amount of effort to track down his precise sources of information. This makes assessing his claims challenging. However his text provides evidence against any supposed 1970s consensus for global cooling. Equally it provides a warning of the effects of failing to think seriously about the politics of climate, a theme I will take up later.

When I first became politically active and concerned about ecological issues, in the early 1980s, climate change was less of a concern for environmentalists than it is today. As the years went by, the risk of rising temperatures because of CO2 emissions became more significant and more commented upon. Incidentally, I remember only one account of global cooling in my first decade of green activism in the 1980s. This was from a book review in the ecological politics magazine *Green Line* of a title by the agronomist John D. Hamaker (1982). The critical review in *Green Line* summarized his argument. At the end of a glaciation the glaciers left behind large amounts of finely ground dust from the rocks that they had crushed, this provided plants were mineral nutrients, boosting productivity. However after a time these minerals from rock dust were depleted, reducing plant productivity and boosting,

> atmospheric concentration of carbon dioxide (this should begin to become familiar now). Resultant climatic changes lead to cooling and increased rain in the polar regions, which causes an extension of the polar ice sheets and a return to glacial conditions. Hamaker believes that our farming techniques (by enhancing nutrient depletion in soils) and fossil fuel burning have hastened this natural and cyclical process: and that by 1990 agricultural production will have been so reduced by global climatic changes that 'civilisation will be wrecked and extinct by 1995'. (Keyes 1985: 16)

Interestingly both Hamaker and Ponte acknowledge the role of CO_2 in climate change. Hamaker, assuming the book review above is accurate, believed rising CO_2 levels lead to falling temperatures (?) and Ponte viewed fossil fuel burning as contradictory in that it both increases CO_2 (pushing temperatures up) and dust (pushing temperatures down via global dimming).

Complex climatology

The rational take from Ponte is the perhaps obvious but nonetheless important point that climatology is a complex and highly contested facet of scientific knowledge. The Earth's climate has fluctuated considerably in the past; a combination of factors have driven changes and there is still some debate as to which factors have been dominant and which have been less significant. As climate changes, rising or falling temperatures give rise to feedback mechanisms. Negative feedback tends to react to cancel out a change in temperature, positive feedback will in contrast reinforce any change. The likelihood of a planet enjoying a reasonable temperature for most forms of life is rather small; in our solar system, at present all other planets are either, as far as we know, far too hot or far too cold to support life. Incidentally, James Lovelock's Gaia hypothesis rests on the assumption that negative feedback mechanisms, evolving over millions of years, have tended, despite fluctuations, to preserve a liveable climate (Lovelock 2007: 45).

One key influence has been the Sun's activity, with scientists debating the importance of solar flares and whether the intensity of our star has remained constant. The faint young Sun paradox still, in fact, remains to be solved. During the first billion years of the Earth's existence, it has been argued, the Sun was thirty per cent less intense than today; this would mean apparently that our planet would have been plunged into icy temperatures. Yet geological evidence implies that the Earth was warmer at this stage than the intensity of the relatively faint Sun would suggest possible (Bengtsson and Hammer 2004).

More sunshine equals higher temperatures but this is

mediated by other factors and feedbacks including the Milankovitch cycles, named after the Croatian born scientist Milutin Milanković(1879-1958). Milankovitch cycles, driven by the Earth's orbit, have been a major influence on past changes in climate. In 2018 the *Proceedings of the National Academy of Sciences* published research analysing a 1,700-foot-long rock core drilled in the Petrified Forest National Park in Arizona. This suggested that the Earth's orbit moves through a 405,000 year cycle, gradually changing from nearly circular to more elliptical over 202,500 years, before changing back over another 202,500 years. Orbital changes occur, 'because of the Earth's gravitational interactions with other planets, especially Jupiter and Venus — Jupiter because it is very large, and Venus because it is very near.' (Bakalar 2018)

While our planet's relationship to the Sun is crucial, as noted earlier, dust in the atmosphere is also significant when it comes to understanding global temperatures. This is one of several ways in which volcanoes can potentially influence climate. Variations in climate have often been hypothesised as the product of volcanic activity, spewing forth both CO_2, other gases and particles. The first person, apparently, to pick up this link was Benjamin Franklin. Writing in 1784 he noted the severity of the winter, arguing that over six months of eruptions from the Laki volcano in Iceland had, by pouring huge quantities of material into the atmosphere, dimmed the light from the Sun, causing plunging temperatures. While there has been some debate, contemporary scientists have produced evidence backing up Franklin's analysis (Burroughs 2003: 88-89). Volcanoes also produce sulphur dioxide, which is one of a number of greenhouse gases. While debates continue, it is clear the climate changed long before humanity made an impact. Cutting through the complexity, temperature changes are overdetermined as they are the product of multiple causes. Of these, the most recent ice ages and interglacials (the warmer periods between ice ages), are seen as likely driven by Milankovitch cycles (Imbrie 2013).

Carbon is king of climate

It has been suggested that not only is human action now the key driver of climate change but that this has been the case for perhaps thousands of years. The palaeoclimatologist William Ruddiman has argued that since the creation of agriculture in prehistoric times, humanity has been clearing forests, often by burning, and creating herds of farm animals. He believes that this has released both methane and carbon dioxide, raising temperatures on a planetary scale (Ruddiman 2003). While climate change through the billions of years of Earth's existence has been complex, the greenhouse effect has been significant throughout. The basic principle is both simple and has been known for over a century. Greenhouse gases limit the amount of sunlight which is reflected away from the Earth and thus heat up the planet, like a greenhouse. Of these greenhouse gases, one of the most potent has been methane, but it is shorter lived in the atmosphere than carbon dioxide. Careful study has shown that during the last century CO_2 emissions have risen, and with them temperatures. Correlations are not explanations; indeed, it is important to understand that not only does CO_2 lead to rising temperatures but that rising temperatures can lead to rising CO_2. This is via several important positive feedback mechanisms. For example, as temperature rise, more permafrost thaws and when it defrosts it leads to further increases in CO_2.

While several factors, as we have noted, influence climate, the scientific consensus is that greenhouse gas emissions have been by far the most significant in recent decades. They don't translate directly into uncontroversial changes in temperatures, but both the logic and the evidence point to rising emissions as the source of rising temperatures. Climate denial as a rhetorical technique often rests on apparently persuasive but easily refuted narratives and tropes. Evidence of rising temperatures has been challenged, but the challenges have been investigated and discarded. While there are a number, as mentioned, of non-human influences that might explain rising temperatures, the correlations are absent. For example, rising temperatures in recent decades are clear but these cannot be linked to an absence of volcanic activity, changes in sunspots or variations in the orbital relationship of the Earth to the Sun. One complicating factor is the existence of negative feedback factors; indeed if feedback factors were entirely positive, we would be in very deep trouble. However, despite efforts to find negative feedback mechanisms that would halt or slow temperature rise, the scientific consensus implies that these are not going to save us from significant climate change. The key negative feedback mechanisms are associated with carbon sinks. When CO_2 is released into the atmosphere much of it is currently absorbed, happily reducing the possibility of warming. Carbon sinks include the oceans and absorption by vegetation. This is not solely good news for at least two reasons. Sinks tend to fill, and the continued absorption capabilities of land and sea are limited. The absorption of CO_2 by the oceans is leading to another problem, that of acidification, which is already damaging plankton and other calcium carbonate based forms of sea life (Malsbury 2020). If sinks absorb less, more of the CO_2 released by burning fossil fuels will be retained in the atmosphere. Increased cloud cover from higher temperatures may have contradictory effects: overall, scientists feel that positive feedback will outweigh negative feedback (Skeptical Science 2011). As temperatures rise, water vapour increases in the atmosphere, this acts as a climate change gas and like CO_2 and methane increases temperatures further, which, of course, in turn, leads to more water vapour. The influence of albedo, noted by Ponte in *The Cooling*, is relatively straightforward. As more ice and snow disappears, more rock and soil, which are darker, are exposed and they absorb more heat, once again feeding back into higher temperatures.

The planet's climate is a complex system, which we will probably never fully understand, and global processes can lead to contradictory regional effects. For example, climate change could affect the Atlantic Gulf Stream, which currently keeps Britain, Ireland and much of Western Europe relatively warm.

If this was disrupted by cold water pouring into the Atlantic as ice in the Arctic melts, temperatures in Western Europe might fall. The worry, on the whole, is that positive feedback will lead to so called tipping points. A tipping point occurs when the climate moves into a new dynamic; a little like a ratchet effect, once temperatures have moved up, it is difficult for them to move back down again. Another concept relevant here is that of nonlinear dynamics. Instead of moving in a steady, linear way, rising temperatures could lead to unpredictable consequences.

Discussing feedback mechanisms (positive and negative), tipping points and nonlinear dynamics, one group of scientists have recently suggested that we are close to reaching an irreversible process of creating what they term 'hot house Earth'. Each year forests, oceans and soils soak up an estimated 4.5 billion tonnes of carbon; without these sinks, all this carbon would enter the atmosphere. Studying ten of these negative feedback mechanisms they note that several of them are close to tipping points, where instead of absorbing they could start emitting carbon. Discussing forests, Arctic sea-ice and methane hydrates in the ocean, it is argued that 'if one of these systems tips over and starts pushing large amounts of carbon dioxide into the atmosphere, the rest could follow like a row of dominoes.'(McGrath 2018b). This would lead to rapid warming.

When we think about climate, we assume that the 'natural' state, is rough stability. While day to day weather can vary markedly, at least in the UK, we feel that there is an approximate and happy state of climate that is good for human life. This assumption ties in with many of the fears around climate change that 'nature' is being disrupted by our collective actions. 'Nature' will react in a violent way with rising temperatures resulting in unpredictable 'climate chaos'. I take a slightly different lesson from all of this: climate is extremely complex, and we can never absolutely predict how it is likely to change. In turn, climate has changed dramatically during the geological history of our planet, even in recent centuries some major variation has been apparent. This perspective has been used by those in denial to say that with climate almost constantly changing in difficult to predict ways, we have nothing to fear. Thus, we can continue to burn coal, oil and gas, to cut down forests and to generally use the environment as we see fit. Reading Fred Pearce's book *The Last Generation*, a couple of years ago, it struck me that, rather than being a piece of harmonious nature that works to sustain human life in the best possible way, the global climate is like some huge beast. Rather than counting ourselves lucky that it is, perhaps temporarily, rubbing along quite nicely with the human species, we are poking and prodding it. According to the climatologist Wally Broecker, we risk rousing an angry and unstable animal; messing with the climate beast is ill advised (Pearce 2006: 16). The vast global experiment of burning stocks of fossil fuels that have built up over millions of years in a matter of decades, looks like a dangerous and unpredictable process that could disrupt a highly complex system. Data from ice cores, using air trapped in tiny bubbles, allows scientists to measure past carbon dioxide levels. They show that carbon dioxide levels are higher now than at any point in the last 800,000 years (Fortuna 2020).

The evidence is clear. Carbon dioxide emissions are rising, along with temperatures. Whatever the complexities, climate change is already with us. NASA noted:

> The planet's average surface temperature has risen about 1.62 degrees Fahrenheit (0.9 degrees Celsius) since the late 19th century, a change driven largely by increased carbon dioxide and other human-made emissions into the atmosphere. Most of the warming occurred in the past 35 years, with the five warmest years on record taking place since 2010. Not only was 2016 the warmest year on record, but eight of the twelve months that make up the year — from January through September, with the exception of June — were the warmest on record for those respective months. (NASA 2019)

The obscenity of climate change is that for short term gain, long term pain will be imposed on both humanity and the rest of nature. Those with *hubris*, a Greek term for pride, face *nemesis*, which translates as retribution. Certainly, while we can never entirely predict future climate, rising temperatures are occurring and the effects of climate change are increasingly being seen as leading to a host of potentially horrifying problems. Lonnie Thompson, an American palaeoclimatologist, famed for his knowledge of glacial geology and ice cores, noted, 'As a result of our inaction, we have three options: mitigation, adaptation, and suffering.' (Thompson 2010: 153)

The climate emergency

One perhaps obvious lesson of climate history is that human extinction is (probably) off the agenda. Human beings have lived as a species through glacial periods and warmer times. The effects of the last glaciation on what is now Britain were remarkable. Temperatures fell and glaciers advanced over what is now an island, wiping much of it down to bed rock. At the peak of the last glaciation, the Devensian, around twenty thousand years ago, a glacial sheet of ice came down to the Midlands, and South Wales, crushing everything in its path (Campbell *et al* 2012: 7). With limited technology Palaeolithic humanity survived catastrophic climate changes, several times over, during cycles of glaciation and interglacials. However, avoidance of extinction is both rather a low bar and subject to some risks that were not apparent in earlier times. Human society is largely settled; our nomadic ancestors would have been able to move from where they lived with less inconvenience. In turn, mentioning things nuclear, stocks of nuclear weapons mean that climate change could lead to devastating conflict. Nuclear power plants, often built on the coast, could also be threatened by rising sea levels.

While there is debate as to how quickly climate changed in the past, the possibility of climate change being measured in decades rather than centuries could make adaption extremely difficult. Equally while the Extinction Rebellion might, if referring to our species, be subject to debate, in regard to other species it is sadly accurate. Species can adapt and move, but sudden changes make this option impossible in some cases.

A report on the security implications of climate change, describes the world in 2050, working on the assumption that emissions stabilise in 2030 and then start to fall. They argue that at least 2.5 centigrade of warming will occur meaning that tipping points for West Antarctic and Greenland Ice Sheets will have been reached. The Arctic will be heading for ice free summers in 2030, there will be significant loss of permafrost and major Amazon dieback. (Spratt and Dunlop 2019: 8). By 2050 many parts of the planet risk becoming unviable for agriculture, flooding will affect many populations, millions of people will be displaced and because of this major social conflict will result. By 2050 large parts of some of the world's most densely populated cities including Chennai, Mumbai, Jakarta, Guangzhou, Tianjin, Hong Kong, Ho Chi Minh City, Shanghai, Lagos, Bangkok and Manila will be abandoned. Ten per cent of Bangladesh will be flooded displacing 15 million people. By this time they believe many of the world's smaller islands will be under water (Spratt and Dunlop 2019: 9).

They note that temperatures may rise more than they predict. Cynics could compare this, and similar reports, to the words of *The Cooling*. Both have a similar feature, prediction of catastrophe from think tankers with military connections. In a few paragraphs, the planet's problems are painted vividly, but the politics is hardly described at all. We can, I guess, prepare for the worst and pray for negative feedback mechanisms to be discovered. The climate contrarians or enthusiasts for market-based solutions or technological fixes, assume that everything will always get better, disregarding regular reports of potential catastrophe.

However, the catastrophe is already with us. The negative effects of climate change and other products of environmental damage are having an effect today. Those who believe that climate concern is climate alarmism, fail to acknowledge effects that are causing harm now. Already, extreme weather events are becoming more common, and while any one of these cannot directly be attributed to rising CO_2 levels, like spots on a canvas that begin to merge together, a picture is apparent. Communities are being displaced by the sea, some are finding it increasingly difficult to farm, permafrost is melting, and populations are moving. Climate change, for example, it has been suggested, is a force driving migration from Central American countries like Honduras and Guatemala, into Mexico and from Mexico into the USA (Gustin and Henninger 2019).

The ramifications of producing fossil fuel-based energy also go beyond the effects, present or future, of climate change. Coal, oil and even gas are dirty fuels with filthy consequences; extraction of these fossil fuels creates harm now irrespective of rising temperatures. The concern used to be that, as finite resources, they would soon run out, but the miracle of the market has incentivised greater exploration and extraction. Yet as new sources of metals and minerals are found, extracted, consumed and eventually burnt or discarded, new wounds are inflicted upon humanity and our natural environment.

It is also important to remember that while fossil fuel extraction is seen as the leading source of environmental damage, it is one amongst many others. From nuclear waste dumping, to the discarding of plastic, to air pollution, environmental problems abound. It is the case, and largely uncontroversial, that thousands of people die prematurely in Britain each year because of poor air quality (Richard 2018). Climate change is part of a web of problems that harm us now, reduce life expectancy and threaten other species. Some problems, such as the apparent rapid decline in insect numbers, are poorly understood. So, in tackling climate change, we should not forget other environmental ills, many of which may interact with climate change to make it even worse. Human history has seen, from the loss of the passenger pigeon to the extermination of the dodo, that careless exploitation leads to extinction.

Finishing this book in December 2019, I have been talking to my aunt in New South Wales, Australia. Much of Australia is literally on fire, New South Wales is one of the worst affected parts of the country. She emails me:

> I'm here for Christmas and New Year ... very warm and has been smoky. No actual fires in Canberra at the moment but the smoke travels from miles away. It's been really bad everywhere. A lot of damage and houses lost in the Port Macquarie areas. The wildlife has fared badly too.... our Koala Hospital has taken in lots of injured koalas.... several hundred killed. Friends have been ready to evacuate. I'm lucky to be in a safe place although it's been too smoky for me to go out on several days. In between the smoke it's been extremely windy when I try to avoid being out. (Memories of being lifted up in a freak gust and smashing wrist). So many houses have been lost, it's so sad everywhere: not much Christmas cheer for lots. (Angela Warren personal communication)

The climate emergency is here, inaction will lead to only more destruction. Adaption is vital, an important topic that would require another book to discuss seriously, but we need to cut carbon emissions. How we do so is the subject of Chapter Three.

Chapter Three
Another Green World

A key attribute of the period was that power did not reside in the hands of those who understood the climate system, but rather in political, economic, and social institutions that had a strong interest in maintaining the use of fossil fuels. Historians have labelled this system the *carbon-combustion complex* (Oreskes and Conway 2014: 36).

This chapter discusses strategies for tackling the climate crisis, looking at the question of lifestyle change to reduce our emissions, notions of a Green New Deal to decarbonise the economy, and moves on to suggest that capitalism needs to be replaced with a new, ecological economy.

How bad are bananas?

Climate change became, for a time but perhaps fleetingly, a big issue in Britain. This may have been due to Extinction Rebellion (whose protests are discussed in a later chapter) and a BBC TV programme on climate change made by the much-loved wildlife broadcaster David Attenborough. In the UK, for a time, everybody or nearly everybody wanted to talk about climate change and what we can do about it. A work colleague struck up a conversation with me saying that he had decided to find out more and had read *How Bad are Bananas?* (Berners-Lee 2011). Subtitled 'The Carbon Footprint of Everything' it is a small encyclopaedia of ecological sin. A range of products and practices are measured against the amount of carbon they produce. It can read a little oddly. I am reminded, scanning the contents, of the Argentine novelist Jorge Luis Borges' taxonomy of imaginary animals. Borges includes fourteen categories. These range from those belonging to the emperor, to 'stray dogs' to 'mermaids' to 'fabulous ones' to 'those that, at a distance, resemble flies' (Borges 1999: 231). Mike Berners-Lee has a simpler system of taxonomy, he classifies by weight of carbon, ranging from under ten grams to over one million kilos. But like Borges' taxonomy of beasts, a host of unusual entries include text messages, bananas (of course), nuclear war, one kilo of trout, universities, volcanoes and cremation.

Since 2009 when *How Bad are Bananas?* was first published, the growth of cyberspace has become an increasing source of emissions. Social media, web searches and crypto currency are causes of growing energy use. 'Mining', the process of creating new bitcoins, generated an astonishing 17.29 megatonnes of CO_2 in 2018 (Vaughan 2019)

Berners-Lee's examples are varied but his analysis is straightforward. The assumption behind *How Bad are Bananas?* is largely that of personal climate arithmetic: if we all consume less carbon, the problem is solved. I am being slightly unfair here, Berners-Lee references Tim Jackson's book *Prosperity without Growth* (Jackson 2017). In doing so he acknowledges that the pursuit of economic growth as a systemic feature of society is a cause of ecological problems. Nonetheless, *How Bad are Bananas?* is symptomatic of an approach to climate crises based on incremental personal action. As such it is, I feel, largely inadequate to solving the problems we face. Berners-Lee notes that the illusionist Derren Brown, 'refers to one of his core techniques as the misdirection of attention: by focusing his audience on something irrelevant, he can make them miss the bits that matter.' (2011: 8). I don't think lifestyle approaches are maliciously constructed to misdirect us from more important practices aimed at tackling climate change, but the fear is that they might have this effect.

While lifestyle action is a common response, perhaps it is more immediate and obvious for most people to cut down on personal climate emissions than to join a political party or protest group, it has been increasingly criticised. Michael Mann, the well-known climate scientist, attacked the kind of response based on *How Bad are Bananas?* Writing with Jonathan Brockopp in *USA Today*, he argued,

> You can't save the climate by going vegan. […] This new obsession with personal action, though promoted by many with the best of intentions, plays into the hands of polluting interests by distracting us from the systemic changes that are needed. (Mann and Brockopp 2019)

Incidentally while there is a section on 'A pint of milk', Berners-Lee does not explicitly discuss veganism. I agree, on the whole, with Mann's analysis and feel I can add some additional criticism of a lifestyle response. Having said this, I don't think lifestyle action is wholly without merit, but to be part of the toolkit of meaningful change, some critical rethinking is necessary.

One problem with lifestyle change is structural. By structures I don't mean factors that structure what we do so as to make change impossible. Nor do I mean structural in the sense of a social theory that argues that reality is the product of language (Sturrock 1979). Instead I would argue that there are factors that shape what we do, making actions easier or more difficult; changing structures is more important than changing personal lifestyles. For example, one of the most important sources of CO_2 is transport; choosing to use cars and planes less is certainly valuable in reducing emissions. However, our personal action works within transportation systems. Many people in London, or other large cities, never bother learning to drive a car, as the density of public transport links, including rail and underground, make it unnecessary. In rural areas, in contrast, it is more difficult to travel without driving. Structural change such as increasing spending on rail and reducing it on motorways, would help people to live in a more climate friendly way.

Likewise, air travel is a major and growing source of emissions. While it is difficult to travel, say, from Moscow to Melbourne without flying, greater investment in rail would make it easier to take the train rather than the plane for shorter distances. Carbon lock-in via infrastructures that promote planes and cars is one example of how individual action is difficult in a context shaped by larger policy decisions. War and militarism are another, perhaps structural source of climate change. The world's armies consume extraordinary amounts of energy:

> America's armed forces use about 30 terawatt-hours of electricity per year—about the same as Ireland—and more than 35m litres of fuel per day. In 2016 a report by the Defence Science Board, a committee of experts, concluded that demand would surge as new power-hungry weapons, like lasers and rail-guns, come to maturity. *(Economist* 2020b)

Another limitation of personal lifestyle action is that it ignores the fact that in a capitalist economy, production is not based on what we want or find useful. Production is for profit and while firms would fail if their products were rejected by us, firms spend a huge amount of money and effort shaping our preferences. The politics of attention and affect need to be addressed if we are to achieve serious social change. Lifestyle change could perversely shape our preferences so that we are more resistant to taking climate change seriously. Given the fact that nearly everything we do, including sending texts, produces some CO_2, and given the complex calculation of our climate impact, discussion of personal change might just discourage us from doing anything.

So far so bad, however I think there is at least one argument in defence of personal lifestyle action. What we do influences what we think, so if we can practice a more ecological lifestyle, we may become more conscious of the need for ecological policies and practices. There is a complex politics of identity and emotional identification bound up with our everyday habits and rituals. If we change our lifestyle, we may become more sympathetic to more fundamental change. I feel this is one reason why environmental groups encourage personal lifestyle action.

Another argument for attending to our personal carbon footprint is that hypocrisy is weaponised by the climate deniers. And they have a point; if a former US Vice-President such as Al Gore is promoting climate action but travelling by Learjet and living in a brightly lit mansion, one wonders how serious he is about the problem of rising carbon emissions. Researching climate change communication, George Marshall looked at the flying habits of climate change scientists. Professor Kevin Anderson astonished his Chinese audience, at a scientific conference, by revealing that he had travelled all the way from Britain by train (Marshall 2014: 201). Anderson is perhaps exceptional; many climate scientists fly on a regular basis. While there are valid justifications for flying, it is one of the most extreme ways of damaging the environment, and what we do affects how seriously or not we are listened to. Marshall references an article entitled 'When Swordfish Conservation Biologists Eat Swordfish'. Its author Giovanni Bearzi suggested that this was 'as if monks advocating poverty were to wear jewelry and expensive silk robes.' (Marshall 2014: 203) It is true, of course, that the army of social media deniers exploit often minor lifestyle lapses or fabricate supposed lapses as a means of questioning any concern with climate change.

Since writing *How Bad are Bananas?* Mike Berners-Lee has refined his analysis and now argues that 'Lifestyle changes are no substitute for collective action'. Noting that climate change action should not be based on misery, he constructively suggests, '*Be kind to yourself.* Very few of us are squeaky clean in carbon terms' but we should try to make progress, 'and *enjoy* the process of cutting carbon out of our lives.' (Berners-Lee 2019)

Lifestyle change should prioritise making life easier and better, rather than acting as a form of environmental puritanism. For some of us, cycling and walking are not options, but if they are, we might enjoy doing so. Of course, campaigning for good safe footpaths and cycleways is vital. Growing some of our own food might be helpful – as I write I am getting strawberries, herbs and kale straight from the garden. Home brewing beer produces less carbon than going to the shops; for a start I reuse the bottles time after time, which eliminates the energy needed to make new bottles. From solar panel installation to better insulation there are ways of cutting carbon that also cut our expenses. However, any lifestyle change risks being moralistic and can become more compensatory than effective. In a capitalist society we are enmeshed in individualism, which is why individualistic lifestyle solutions often, in my view wrongly, dominate discussion of climate change action.

Good technology

We could, of course, bypass the tedious counting of carbon by inventing a machine to suck up the world's excess carbon dioxide. It might be argued that we already have such machines: they are known as 'trees'. There are a wide range of supposedly good technologies that might be used to overcome the threat of climate change and other ecological ills. Indeed, looking at the BBC as I write, it has an enthusiastic news item about a Mexican scientist who has discovered a process for turning cactus juice into biodegradable wrappers that can replace plastic (Heyden 2019). The *Economist* relates that the Environmental Defence Fund is launching MethaneSAT in 2022. This satellite can scan the planet, picking up methane concentrations 'as low as two parts-per-billion' (*Economist* 2020a). Spotting and stopping accidental releases of methane from pipelines is a technological solution that has the potential to reduce emissions of a significant greenhouse gas.

Technological approaches to environmental protection are proliferating. It has been suggested, perhaps most radically, that human beings could be biomedically altered so that future generations of humanity might generate fewer carbon emissions. The 'biomedical modification of humans to make them better at mitigating climate change' might include making us physically smaller so we use less energy, introducing meat intolerance and making us more empathetic to others using pharmaceutical interventions (Liao *et al* 2012: 207).

Another technological approach is via geoengineering (Hulme 2014). Aware that temperatures are already rising and scary feedback mechanisms kicking in, advocates of geoengineering argue that only large-scale technical solutions can save us from disaster. The aim is to manipulate global climate. Humanity (or at least some of us) could use a planetary thermostat to consciously adjust temperatures. One challenge to geoengineering solutions is the global complexity of climate, which could lead to sharply negative consequences. The most commonly discussed solar radiation measure, to reduce heating, would involve injecting sulphate into the stratosphere. One possible result of this could be significant depletion of the ozone layer and it might damage rainforests in some parts of the world (Burns 2013: 203)

Choices would have to be made as to winners and losers from such global climate manipulation. It could have a major impact on agriculture in particular countries and regions. Even if geoengineering were possible and desirable, there are major technological and funding issues to be resolved, while its implementation would have social and economic consequences. This points towards the two major problems with using technological approaches to climate change and other ecological problems: First, in a world based on capitalism, economic accumulation is the universal goal, and this tends to degrade nature. Thus, providing technological solutions without challenging the growth imperative tends, at best, to delay destruction rather than overcoming it. Secondly, technological choices are political choices. Technology is never neutral, the introduction of a technology creates winners and losers, empowers some and potentially disempowers others. Technological solutions may be part of the process of gaining a more sustainable world, but they don't stand alone: they shape and are shaped by political, economic and cultural processes.

Geoengineering or human biological transformation might be rejected, and a simple advocacy of renewable energy might appear in contrast largely uncontroversial. Renewable energy, however, is not without costs and limitations. For, example, advocates of fossil fuels point out that energy generated by sunshine, waves and wind is not continuous but depends on the right kind of weather. Iconic propaganda threatens us with power outages, and higher bills, if we reject coal. However, with modest technological improvements renewable energy has become ever cheaper. Equally, while the intermittent nature of many renewables is a problem that needs to be dealt with carefully, it is less drastic than might be thought. A diversity of different renewables is helpful; if the Sun doesn't shine the wind might blow. Power supplies are increasingly interconnected internationally, which makes renewables more dependable in the face of supply fluctuations (Liu 2015). Batteries and other forms of storage are also advancing (Delbert 2020).

There are a number of other challenges. Hydroelectricity is a major and established source of renewable energy; Venezuela, for example, has an economy based on some of the world's dirtiest oil, but its domestic energy needs are largely, at 64 percent of total supply, met by hydroelectric (World Bank 2015). Yet climate change, by creating more extreme weather patterns, may accelerate droughts; reduced rainfall is already making some hydroelectric generation fail in various parts of the world. A future of climate chaos will make hydro far less reliable (Holthaus 2015).

The Jevons paradox is another challenge (Foster *et al* 2011: 179). Named after the nineteenth-century coal economist it shows that energy conservation can paradoxically lead to more energy consumption. As more of us conserve energy, demand for energy falls. This pushes down prices and, in turn, encourages greater consumption. One manifestation of this is car use. As cars have become more efficient in recent decades, they use less fuel. This has the potential to make driving cheaper and thus increase the number of cars on the road.

The British government has promised to eventually phase out petrol- and diesel-powered cars. Transport is a major contributor to carbon dioxide emissions, so plans to make sure that all cars are electric appear positive. However, while welcoming the government's decision to reduce emissions, a number of geologists have questioned the practicality of electrifying transport. They suggest that electrifying cars and vans in Britain by 2050 would require '207,900 tonnes cobalt, 264,600 tonnes of lithium carbonate, at least 7,200 tonnes of neodymium and dysprosium, in addition to 2,362,500 tonnes copper'. They calculate this would be nearly twice current annual world production of cobalt, three quarters of lithium production and nearly fifty percent of copper production in 2018 (Natural History Museum 2019). Scaling this up to the world's vehicle fleet would produce astonishing numbers. Of course these figures can be challenged, and shifting to electric is more desirable than burning oil, however business as usual with a battery may be logistically impossible.

Renewables also create some environmental and social costs. Proponents of a high technology 'green' future talk of building solar panels in the world's deserts and connecting up virtually unlimited clean energy. However, deserts are not empty but have a diversity of plants, animals and peoples who might be disrupted by blanket development (Komoto and Kurokawa 2008). Likewise, the big dams to make hydroelectric work, if they have a future in a world with less predictable weather patterns, leave a political footprint. The valleys flooded for 'clean' energy displace the people living in them. Minorities often suffer; in the UK, Wales has been the site of large reservoirs, while in Turkey, huge dams have been built in Kurdish areas. Nick Estes, assistant professor of American studies at the University of New Mexico, and a citizen of the Lower Brule Sioux Tribe, has catalogued how dam building displaced the Lakota and Dakota nations during the twentieth century. He notes the US has used the power of eminent domain to take Indigenous land for dams,

All of the risks, and none of the rewards, of cheap hydroelectricity and irrigation, were imposed on generations of Indigenous people who depended upon their relation to the land and water for life. And floodwaters provided the physical means to terminate Indigenous nations and relocate people – a violent severing of those relations – to end the 'Indian problem' once and for all. (Estes 2019: 133-134)

Biofuels are another example of a potentially costly form of renewable energy. Growing crops to turn into fuel sounds attractive. One is reminded of the scene in the 1949 British Ealing comedy film *Whisky Galore!*, where, running out of fuel and being chased by the authorities, the scavengers of the whisky from the wrecked ship *SS Cabinet Minister* smash open a bottle and pour the contents directly into the tank. They make good their getaway. What could be greener than growing fuel? Yet, as has been now well documented, biofuels are highly problematic. Growing crops for fuel may displace crops used to feed people, perhaps pushing up prices and increasing hunger. Agriculture is also a major source of carbon emissions, using petroleum-based pesticides and fertilizers may mean that growing such energy crops increases, rather than reduces, emissions. Energy crops may also dislodge wildlife and even marginalise human populations. In Colombia, a major source of palm oil imported by the European Union (EU), right-wing paramilitaries have been known to threaten local communities and steal their land to grow this type of biofuel (Wall 2009). Of course, the violence meted out to local people, in order to enclose territory and shift populations for coal and oil, continues the world over. For example, the novelist, essayist and war correspondent William T. Vollmann's *Carbon Ideologies* examined resistance to an open cast coal mine in Bangladesh (Vollmann 2019b: 256).

Yet another technological approach to climate change is carbon capture and storage (CCS) (Stephenson 2013). The idea is that if carbon dioxide can be taken from the atmosphere or scrubbed out of the emissions from coal, oil or gas fired power stations, the problem will be solved. It is a worry that if carbon is captured, it might be released at some point in the future, risking higher temperatures. Perhaps clean coal is possible, but the more obvious assumption is that even without carbon emissions, coal will remain a highly invasive and polluting fuel. Carbon capture will be used to justify more coal extraction, and while making it look more acceptable to produce fossil fuels, carbon dioxide levels will continue to rise. The environmental researcher and activist George Marshall fears that CCS may act as 'less of a real solution than a much funded narrative ploy'. He argues that as such it provides a justification that helps to excuse the continued production of fossil fuels (Marshall 2014: 179).

While technical problems and other costs are common to any purely technological attempt to solve climate change, we should not reject technological solutions automatically. Yet seeing them as politically neutral and inevitably desirable invites disaster. Equally, we should not fall for the nirvana fallacy. Nirvana, the Buddhist state of perfection, is also the place, apparently, of nonexistence. There is a danger that in seeking solutions that have no costs, risks, or potential problems that we instead do nothing. Valid criticism of technological alternatives to fossil fuels can be mobilised by advocates of corporate carbon to justify sticking with coal, natural gas or oil. The problem is that we are not engaging in rational if sometimes abstract debate. Instead, in discussing climate change, as in other matters, we exist in a particular context that shapes what we think, discuss and decide. Power structures are apparent in any society so there is a danger, through conscious action or unconscious bias, that whatever we do we end up with alternatives that best fit the needs of the rich and powerful. The most vital point is political: how do we raise the forces necessary to make sure that any policy change actually reduces CO_2 emissions, and does so in ways that also best promote democracy, diversity, equality and justice? This, of course, is the central concern of this book.

A Green New Deal

The Green New Deal (GND) is an increasingly popular response to the climate crisis. This imagines a huge investment into clean energy technologies, public transport schemes and other major projects. The aim is that such an injection of cash will increase economic growth while providing a transition to an economy that works without raising CO_2 levels. GND is a proposal based on the contention that climate change is so severe a threat that government action on a massive scale is the only acceptable response. Personal or local action is inadequate, neither is a carbon tax or any similar method of increasing the price of carbon sufficient. A mammoth technological effort based on transforming the infrastructure of an entire country or countries is essential. Originating in Britain in 2008, the GND is also a Keynesian policy aimed at increasing economic activity. The economist John Maynard Keynes in his *General Theory* written in the 1930s suggested that, left to the market, an economy would move through a cycle of boom and bust. When the economy was 'bust', and growth was low or negative, economic pessimism would gain hold of both consumers and business, discouraging spending and investment. Keynes, to vulgarise and simplify his nuanced and complex attempt to preserve global capitalism, called for governments to increase their spending and borrow during periods of negative growth to kick-start the economy (Keynes 1936).

While working without knowledge of Keynes' theory, the US President Franklin Delano Roosevelt embarked on the New Deal in 1933. This involved three strands: massive public spending on infrastructure, including big dams; reform of the banking system: and increased welfare to help the unemployed and otherwise vulnerable. It is generally credited with reviving the economy from the 1930s depression. The GND is named explicitly after Roosevelt's project, and also references the economic planning needed by the allied forces in the Second World War. The original GND report published in Britain in 2008 challenges free market and piecemeal alternatives:

In our living memory, the scale of economic re-engineering needed to prevent catastrophic climate change has only been witnessed in a wide range of countries during war time. No other approach looks remotely capable of delivering the necessary volume of emissions reductions in the time needed. (Simms *et al* 2008: 41)

The authors of the 2008 report argued that climate change, rising oil prices and financial crisis meant that a bold solution was needed. Recently, the GND has been advocated by figures on the left of US Democratic Party such as Alexandria Ocasio-Cortez and Bernie Sanders (Aronoff *et al* 2019: 15). Both the US and British versions have stressed a planned, technological approach to climate change, associated with a left Keynesianism that promises greater regulation and greater equality. Investment in renewable energy, home insulation and public transport are all elements of the project. In the US, the forces of Trump and the right have ridiculed the project but there have also been critics on the left who see its apparent ambition as insufficient to the crisis we face. In turn, while broadly supporting its implementation, the environmental anthropologist Myles Lennon has argued that the GND can have problematic effects. He contends that decarbonization and 'energy transitions are infinitely more complex than the rhetoric of "100 percent renewable energy" and "millions of green jobs" suggests' (Lennon 2019). He notes that supposedly 'green technologies' may be exploited by financial markets to make speculative profit, may rely on exploitative labour conditions and, may, indeed, have a significant carbon footprint.

Perhaps the greatest problem with the GND is its Keynesian inspiration which has as much to do with preserving capitalism as conserving ecosystems. As such it exists not to challenge the logic of capitalism, but to protect and nurture capitalism. It will have a positive effect, and provides some real gains, but like other broadly technological solutions to climate crisis, it ultimately fails to move us beyond the current economic system.

In the end we have to move, I would argue, to an ecological economy that rejects continuous growth and ever increasing capital accumulation.

It's the economy, stupid

Another aspect of the debate over technological solutions to climate change has been the critique of ecomodernism. Ecomodernists argue that technological development is accelerating and that innovation can solve the climate crisis. Changes in lifestyle and culture may be unnecessary because new energy sources, genetic engineering and artificial intelligence can create a low carbon future. This has been criticised, most stridently perhaps by advocates of 'degrowth' (Symons 2019b). 'Degrowth' has been defined as 'a downscaling of production and consumption that increases human well-being and enhances ecological conditions and equity on the planet.' (Research & Degrowth 2019) Degrowthers argue, in summary, that increases in economic growth tend to degrade the environment. For example, economic growth tends to increase CO_2 emissions. Any technical fix is ineffective because increasing economic growth tends to promote more energy use and, perhaps, any form of energy use has some negative environmental and indeed social effects.

At present, nearly all countries pursue a central goal of increasing Gross Domestic Product (GDP), or similar metrics. It is symptomatic of this that attempts to deal with climate change are seen, by policy makers, even those who are sympathetic to environmental goals, as having to fit around the central objective of continual economic growth. The political economy of growth is based on an ideology that more material consumption is desirable. It is virtually a heresy to question this. The idea that growth should be a means to improve wellbeing has been overtaken, perhaps, by the notion that growth rather than wellbeing is the key objective to pursue.

In contrast I would suggest that prosperity can potentially be decoupled from economic growth. We can, as Tim Jackson argues, enjoy higher material welfare with reduced economic growth (Jackson 2017). The paradox is that a growing economy grows when we throw things away and replace them. Social sharing, for example, in a variety of forms from tool and toy libraries to effective public transport, means we can have more access to what we desire with less consumption, production and waste. The point to grasp is that if we consume goods because of the welfare they provide us, there may be ways to provide the welfare with less consumption. Thus, we use a car to get from A to B, but an effective bus, car pool or even cycle paths would reduce the need for as much car use. Maybe with the internet we can avoid the journey altogether, perhaps saving more time and emissions. In short, material goods and immaterial services provide benefits, but the benefits could in some, perhaps many, circumstances, be provided with less production, consumption and economic growth (Wall 2008). Elinor Ostrom's work on the commons suggests that, with good practices and institutions, we can manage resources in a more sustainable and environmentally beneficial way (Ostrom 1990).

Equally, if goods are made to last longer, less growth is necessary. An economics of repair is an element of prosperity without growth. We can and should dethrone growth as the central demand of our society. Even if growth was environmentally sustainable, we need more nuanced ways of seeking to increase economic welfare. There are some signs that growth is now being questioned. Recently the Aotearoa New Zealand government has advocated replacing economic growth with a set of alternative goals: 'thriving in the digital age; improving mental health services; reducing child poverty; developing a low-emission, sustainable economy; and addressing inequality, especially among the country's Maori and Pacific Island peoples.' (Dickinson 2019)

However, while I think it is possible to provide increasing economic well-being without ever increasing growth, economic growth remains essential to the capitalist economic system. A capitalist society is a mode of production that promotes the pursuit of profit. In doing so it has followed a logic of extraction. Taking resources, turning them into commodities (products exchanged for money), selling these commodities and encouraging us to consume and throw them away. The imperative to accumulate profit and capital is structural. It is never a matter of good or bad, greedy or generous capitalists, instead it is a matter for capitalists of survival. Marx argued, to simplify a complex discussion, that early adopters of technology tended to prosper. Growth is necessary for profit, profit is necessary to reinvest, those firms who fail to grow, to make profit and to invest, tend to get wiped out by those who do. Marx bluntly stated, 'Accumulate, accumulate, that is Moses and the Prophets! […] Accumulation for the sake of accumulation, production for the sake of production' is the rule in a capitalist society (Marx 1977: 742). This basic logic acts as a forcefield that shapes nearly everything we do. Thus, as the Marxist literary theorist Frederic Jameson noted, it is perhaps 'easier to imagine the end of the world than the end of capitalism' (Jameson 2005:199).

There are other functional arguments as to why economic growth is necessary to the workings of a capitalist system, including the need to repay debt. Growth reduces the sting of inequality and conserves an unequal system. The concept of a mode of production, or social formation, that we gain from Marx suggests that material considerations, including economics, have a lot of weight in explaining how a society works. The mode of production can however be seen as 'overdetermined', this concept from Freud (2010) is, I would argue, also apparent in Marx. Rather than reducing an event to one cause, causes are multiple. Marx was, in this respect, an ecological thinker, not just in the sense that he was concerned with environmental issues, but in that his thinking was based on networks, contradictions and relationships. His approach was shaped by an ecological philosophy or methodology. For example, his key categories such as 'labour', 'capital', the 'commodity' are not defined as 'things', but their nature is established in relationships (Foster 2000). Ecology, of course, is a science of relationships, relationships between species, pointing to the philosophical ecology of relationships used by Marx, Spinoza and Hegel. I digress.

An economy isn't just money, but might be seen as an interlocking system. Any species produces, humans are no different, but what and how we produce is a complex mode, a way of being. The narrowly defined 'economy' is a product of culture, psychology, politics, etc, etc. It might be seen as a whole way of life. While societies are subject to change, often in the form of sudden breaks, they can have astonishing powers of conservation and may tend to move in a particular direction. While human societies are diverse, there is unity in diversity, we all seem to march to a particular economic rhythm and the difficult fact is that, at present, this seems to be a rhythm of destruction. It is not enough to interrogate our bananas as to their contribution to the carbon budget but instead we must question an entire way of life.

Growth is a deeply embedded feature of our current capitalist mode of production. This is why solutions to climate change, such as the Green New Deal, generally seek to maintain economic growth. Yet challenging the growth imperative remains vital to dealing with the climate crisis. So, if the economy is a key element in our society, with growth buttressed by political and cultural factors, confronting growth will mean moving to an alternative mode of production. This is extremely challenging. It's not so much that we should be discussing individual climate change emissions but instead that we should reflect on the more difficult fact that our economic system is dysfunctional. Or to put it in another way, the way it functions tends to degrade ecological systems and in doing so threatens both humanity and the rest of nature. The threat includes climate change but goes far beyond it.

Time is desperately short. Indeed we may have already tripped through various tipping points that mean we are on the way to hothouse Earth. Therefore, we must limit emissions to prevent further climate change and adapt to the unfolding catastrophe, while seeking to change our destructive mode of production. Too often these strands of adaption, reform and revolutionary change are promoted rather abstractly; groups of activists or academics talking to each other, often engaging in fierce battles on paper or social media. It is important to focus on the material effects we need. The danger is that the denialist right are accelerating climate change, while the dominant ways of contesting climate change, even when implemented, are ineffective, and adaption, where it is being promoted, is on the terms of the rich and powerful.

We need to focus in dealing with climate change to build capacity for necessary change. In the next chapter I will take a look at the success and failures of Green Parties in their attempts to create a green future. To what extent do Green Parties globally provide an effective strategic response to climate change? Have they built capacity? The ambition is to encourage Green Party activists to reflect on such questions and to challenge them to raise their game, as we all should do in the face of an unfolding and diverse climate crisis.

Chapter Four
Green Parties and Climate Change

A commons approach is vital because it is the only way that we can solve the world's problems. Most importantly the climate emergency. The argument that everything we need to survive should be provided in a free and universal service is one that is easy to make, and places rights, not money, at the heart of the argument.

It took us three attempts to get free public transport adopted by the Scottish Green Party at conference. That's one of the joys of having a democratic party. It also demonstrates a move in wider thinking away from the market solutions favoured in the 1990s to environmental and social problems, and towards a commons approach. [...] free public transport was just the first step (McColl 2020).

It is May 2016, I am sat in a hotel bar in Utrecht, waiting for my bus back to Victoria station, London. I had attended one of the European Green Party (EGP) congresses, which take place twice a year. The EGP is, as the name suggests, a body that represents Green Parties from Portugal to Poland. At the time I was International Coordinator of the Green Party (of England and Wales), a job share with a colleague. I strike up conversation with another Green delegate. He is also an International Secretary representing an Eastern European Green Party. Long before I took on an international role, he was regularly attending Green Party conferences in Britain, so he was a familiar figure. I asked him what he did for work. He became, if I remember rightly, a little quiet, then he told me he was an air traffic controller. That didn't worry me, we all have to earn a living, and how we do so might not always fit with our political commitments. Indeed, plans and projects for conversion are important, you can't just dump people into unemployment if their careers are ended because of climate policies. He seemed reassured and decided to tell me about Heathrow Airport, which he insisted needed to expand. He was enthusiastic about plans to build a third runway. Challenging him on this, he insisted it was the green thing to do, as planes were spending too much time waiting to land, circling over London and adding to pollution. This was not even the strangest experience I had in the International role. Many Green Parties do an excellent job but just because they are called Green, it doesn't mean that their contents always fit with the label.

In this chapter I will discuss the extent to which Green Parties are effective in promoting a greener future including their response to climate change. They look likely to be a permanent part of the political landscape in much of the world. While their fortunes have and will no doubt continue to fluctuate, currently they are making some progress. The European Elections of 2019 saw them win additional Members of Parliament, in Britain increasing from three to seven. Opinion polls in June 2019, show the German Greens remarkably in first place, and there is evidence of a long-term trend that they are displacing the Social Democratic Party (SPD) as the main opposition to the centre right Christian Democrats (CDU). Greens have also made recent gains in Ireland, Norway, the Netherlands, Finland, France and Switzerland. The pattern in Europe is not universal; Green Parties are generally weak in Eastern Europe and seem to have collapsed in Italy.

Greens Parties are less successful outside of Europe. They are well established in Canada and the USA, but success is limited by the electoral system; the two-party nature of US politics is particularly brutal. The Canadian Greens, despite the lack of proportionality, have been able to elect members of parliament.

The Green Party of Aotearoa New Zealand are currently in coalition government. The Australian Greens have managed to elect members across various levels of governance. While Africa has many Green Parties; South Africa is a significant exception, this may be as result of the dominance of the African National Congress, although without further study I am not able to say. African Green Parties are generally small. Latin America has seen some surprisingly successful Green Parties including those in Mexico, Brazil and Colombia. In both Brazil and Colombia, Green Presidential candidates have come close to victory on several occasions. The green travelogue includes Asia, with Green Parties currently active in the Philippines and Taiwan but absent in India and Japan. Green Parties have also some presence in the Middle East. In Turkey, the Green Left is an official partner of the EGP and has worked in coalition with the largely Kurdish Peoples' Democratic Party (HDP).

This brief survey, based on personal experience rather than academic sources, is descriptive rather than analytical. To understand the potential of Green Parties, local context is important, and it is also useful to briefly outline the historical evolution of Green Party politics. While it is not a simple task to outline Green Party ideology, I have tended to use the four pillars of the 1983 German Green manifesto as a useful shorthand. The four pillars comprise ecology, social justice, grassroots democracy and nonviolence (Die Grünen 1983). All could be debated: social justice has been a subject of endless academic and political controversary, for example, for hundreds of years. In recent decades, the likes of John Rawls, Robert Nozick and Amartya Sen have exchanged perhaps millions of words debating social justice. Concepts such as 'grassroots democracy' are equally contested. Nonetheless, Green Parties have tended to combine environmental concern with commitments to equality, a less violent international order and the decentralisation of power.

The first Ecology Parties originated in the early 1970s and were more strictly focused on environmental issues. The current Green Party of England and Wales was established in 1973. Initially known simply as 'People', it became the Ecology Party in 1975, the Green Party in 1985, and the Scottish Greens split away in 1990. The Greens in the north of Ireland, once part of the Ecology Party, are now a regional party within the Irish Greens. Other early parties include Values, which later evolved into the Aotearoa New Zealand Green Party. The political orientation of such ecology parties was in some respects conservative; founding members were often drawn from the right. The founding assumption was that both economic and population growth were driving an ecological crisis, although climate change was ignored. Sceptics might challenge greens for producing a continuous litany of catastrophe.

The strategic orientation was relatively clear. Ecological crisis was more important than any other political issue, a plan for a sustainable society involving major transformation was necessary. This was legitimated by science and indeed *Blueprint for Survival*, the key document behind the foundation of what is now the Green Party of England and Wales, was endorsed by many scientists (Goldsmith 1972: 15). While much of the science has been debated and challenged, the foundation of green politics was an argument that infinite growth was unsustainable. The authors of *Blueprint for Survival* aimed to convince government – or failing this – to form a Movement for Survival, which would contest and win a General Election, bringing in a planned transition to an ecological society. While environmental concerns were high on the agenda in the early 1970s and oil shortages added to a mood that might see business as usual as impossible, the Ecology Party had a limited impact.

Green Parties took off in mainland Europe during the 1980s. There are complex reasons for their relative success compared to the Green Party of England and Wales, but two factors are, I think, important. One is the electoral system: many countries used proportional representation, so with as little as five per cent of the vote, some parliamentarians could be elected. A second factor was that Green Parties emerged out of social movements,

movements against nuclear power and nuclear weapons. In doing so they drew in support from diverse communities. The children of 1968 who wanted a more radical but more countercultural left helped provide a generation of initial supporters for the Greens in countries like France and Germany.

While there are good Marxist critiques of the Greens, much of the growth of the Greens was fuelled by disillusionment with the traditional left, both reformist and revolutionary. This was less of a feature of the Green Parties in the UK, at least before 2000, but in Germany, for example, ex-Maoists flocked to the party (Hülsberg 1988). The 1980s saw an important strategic debate amongst European Green Parties, with considerable conflict between 'realos' and 'fundis'. In Germany, the realists urged coalition with the Social Democratic Party both at regional and state level. In contrast, the fundamentalists argued that the ecological crisis demanded an approach that was so radical that working with other parties would lead to dilution. Rudolf Bahro, a former East German dissident, whose politics moved from critical Marxism to green fundamentalism, posed the debate in terms of a choice between ecological revolution and damaging reformism. He argued, to paraphrase a little, that reforms were like freshening the dragon's breath. Instead, he argued, we need to kill the dragon rather than killing the source of the halitosis. Increased investment in renewable energy etc. would take the pressure off the system enough to maintain the system; reform would prevent the revolutionary change needed for fundamental transformation (Bahro 1986: 184). Incidentally, while gradually emerging as an issue, climate change was not a key concern of either 'realos' or 'fundis' during the 1980s.

In the Britain 'realos' and 'fundis' fought bitterly over issues of internal organisation and party structure. The Ecology and later Green Party was hardly in a position to be able to enter coalitions since they polled too few votes to achieve electoral success. The 'realos' argued that a more conventional party structure was necessary; the fundamentalists argued that political realism would lead to capture by a bland elite. The 'fundis' were generally influenced by a more anarchist conception of politics, seeing the party as part of a broader green movement, emphasising the importance of non-violent direct action rather than purely electoral politics (Doherty 1992).

Several continental European Green Parties came from the left. For example, the Dutch Green Left (Groen Links), was formed by a merger of the Communist Party, Radicals, the leftwing Evangelical People's Party and Pacifist Socialist Party. In a UK context, back in the 1980s, this horrified those on the 'realo' wing, who believed in 'a neither left nor right' green politics. However what perhaps was not foreseen, particularly by those who feared the left, is that instead of turning the Greens into ecosocialists, the entry of the left was an episode in their narrative of moving to the centre. The Dutch Green Left, along with the generation of ex-Maoists in the German Greens, grew older, and more moderate. The young student activists of the 1960s like Danny Cohn-Bendit, notorious for their anarchistic energy, became besuited moderate parliamentary Green Party politicians by the 1990s.

The radical 'fundis' appeared, in most European Green Parties, to have dropped out of involvement by the 1990s. The social ecology of Murray Bookchin, the somewhat curmudgeonly New York anarchist, seemed to have been a significant inspiration for German 'fundis' like Jutta Ditfurth (Biehl 2015: 242). The USA Greens were also for a time influenced by Bookchin and were, initially, sceptical about establishing themselves as a primarily electoral party at a national level. The 'realo' versus 'fundi' debated played out differently in the Green Party of England and Wales. I was close enough to this to have to admit bias, yet while party structures were changed, the party in some ways became more radical. A top down managerial style of political practice never quite took hold, and pursuing a purely electoralist strategy proved unsuccessful in the 1990s. Proportional representation, introduced for some elections including the European Parliament, the Greater London Assembly and for Scotland, improved electoral fortunes. The steady movement of the Labour Party away from the left, under Neil Kinnock and Tony Blair, meant the Green Party picked up former Labour Party members. Clear electoral space on the left encouraged the Green Party of England and Wales to shift ideologically. Organised ecosocialists also had the effect of producing a party more clearly on the left.

All these things are contingent. Some European Green Parties seem in recent years to have moved a little more to the left, British politics as I write has been dominated by Brexit, and the Greens have shifted from mild Euro scepticism to full on enthusiasm for the EU. In doing so they have recently gained members and votes. With Britain exiting the EU and a Conservative government in power since the 2019 General Election, it is difficult to predict the political fortune or direction of the party over the next decade. The Greens in the North of Ireland and Scottish Green Party seem to be progressing; proportionality gives them seats and changing attitudes have opened up political space. Their prospects would require much more detailed analysis than I have room for here.

Greens gone wrong?

A number of criticisms have been made of Green Parties. The conservative origins of the Green Party of England and Wales might be challenged from the left. Some forms of environmentalism have come from the right, leading advocates of early ecological politics such as Teddy Goldsmith were previously members of the Conservative Party, and Goldsmith developed a right-wing ecological politics (Barberis 2000: 26). However, origins perhaps don't tell us too much. That group 'X' started with ideology 'Y' is far from proof that they won't change to position 'Z' in the future. Indeed, Green Parties have been quite varied from country to country and have seen changes in ideology and policy over time.

Many Marxists and anarchists believe that Green Parties are perhaps inevitably flawed organisations. On the left, some have pointed to Greens' lack of class analysis and connection with trade unions and other working-class organisations. Many European

Green Parties, for example the Irish, German and Austrian Parties, have also been criticised for making damaging political compromises (Laker-Mansfield 2015). Murray Bookchin, from an anarchist perspective, argued that participation in national elections leads to catastrophe. He felt that reform efforts could, at best, slow 'the overwhelming momentum' towards ecological disaster, arguing that Greens 'seeking state power […] have generally attained little more than public attention for their self-serving parliamentary deputies and achieved very little to arrest environmental decay.' (Bookchin, in Bookchin and Foreman 1991: 77)

Bookchin, with his usual bluntness, certainly identified a problem, but was I feel less successful in providing a solution. The German Greens, despite (or even because of) their electoral success, have been subject to much criticism on the left and beyond. They have been variously described as 'neoliberals with wind farms' (Nagel 2017) or 'neoliberals with bicycles' (Humphrys 2012). From in the early 1980s being perhaps the most radical Green political force, they eventually became a broadly centrist party and a stable part of the German political scene. They have generally been supportive of at best social democratic reforms and on occasions have even gone into regional coalition governments with the conservative CDU. Strong advocacy of the EU has also been a feature of their approach.

The Irish Greens were widely censured by the left and others, including former members, for their coalition with Fianna Fáil in 2007. Irish politics is dominated by two centre-right political parties, Fianna Fáil and Fine Gael, who trace their origins back to divisions within the Republican movement during the Irish Civil War of 1922-23. At the time of the 2007 General Election, one Green Party TD (member of parliament), Ciarán Cuffe noted 'Let's be clear. A deal with Fianna Fáil would be a deal with the devil. We would be spat out after five years, and decimated as a Party.' (Cuffe 2007) This proved prophetic. The Greens went into coalition with Fianna Fáil, the financial crisis hit, and policies of bank bail outs and austerity supported by the Greens led them to lose all their TDs in the subsequent 2011 election. I am not convinced that they have learnt much from this, but in 2019 they polled strongly. Indeed they gained the highest first preference vote of any party in the Dublin constituency in the European Elections, electing Ciarán Cuffe (Fegan 2020). I don't think the Irish Greens' experience is necessarily evidence of a wider failing of Green parties, although it might have been wiser to stay out of coalition. European Green Parties have often gone into governing coalitions and thereby achieved some useful policy gains, but the wholesale critique of growth has, partly as a consequence, generally been lost. In 2020, just on the cusp of finishing this book, the Austrian Green Party went into coalition with the conservative People's Party. Agreement on climate change policies such as higher taxes on flying, cheaper rail fares and a commitment to zero carbon for the country by 2040, might be seen as positive. In exchange, the Party has opted to support restrictions on migration and a ban on headscarves for school girls (Bell 2020). Can we imagine Greens going into coalition with Trump? Probably not, but it is alarming when Greens align with Islamophobic right-wing forces, as in Austria.

When Greens are criticised, the Irish and German examples tend to get mentioned. However, there are more problematic examples of Green Parties than these. Perhaps any international movement family will contain some uncles we would rather not meet or cousins whose behaviour embarrasses us. While there is a Global Greens organisation (https://www.globalgreens.org), members may have surprisingly little or no actual commonality. It is largely the case that individuals can establish Green Parties, and in some parts of the world these apparently exist with little or no scrutiny. The EGP, made up of national parties across the continent, undertakes research into potential members. However, my experience, as noted at the head of this chapter, is that the descriptor 'Green' doesn't always mean very much.

The Mexican Green Party (*Partido Verde Ecologista de México*), I think it is fair to say, would probably be extremely competitive in any contest to establish the worst Green Party in the world. It supports fracking, 'plies peasant farmers with herbicides' to gain votes, and 'in short, acts in ways that are anything but green.' (Agren 2015) The Senator Jorge Emilio González Martínez, known as the El Niño Verde (the green boy), was videoed in 2004 apparently negotiating a bribe to build a hotel in Cancun on environmentally sensitive land. 2011 saw a Bulgarian model fall to her death from a penthouse the media said belonged to him. Other elected officials of the party have been associated with a string of scandals (Agren 2015).

The Mexican Greens were thrown out of the Global Greens when the EGP objected to their support for the reintroduction of the death penalty in Mexico. At some point, and I have not been able to find out why, they were readmitted. It has been argued that they may have been set up by the then dominant Mexican Institutional Revolutionary Party (PRI), as a way of taking opposition votes. Generally, they have gone into coalition with the PRI. Their policies have often seemed far from environmentally friendly. They have enjoyed some electoral success and from this are well funded. It has been suggested that they in turn donate to smaller Latin America Green Parties, who in turn support them. The US and Canadian Greens seem to have done little to challenge their influence in the Federation of Latin American Green Parties. Green Parties, internationally, without being unduly proscriptive, should be more careful which organisations are credited with the label and allowed to join or indeed remain in the family.

Getting to Green

The example of the Mexican Green Party is egregious. Yet while the global Green Party family includes the Mexicans it makes it difficult for Greens to justify their criticism of others in moral or ethical terms. It illustrates the essential principle that in politics, effects matter; Green needs to be measured, not by name or institutional affiliation, but by what such organisations do. However, there are plenty of positive examples of Green Parties working to deal with climate change. In London, Green

Party Assembly Members have pushed over decades to improve cycling, walking and public transport facilities. In Aotearoa New Zealand, the Green-Labour coalition government established a climate commission and banned the KiwiSaver scheme, in which workers are automatically enrolled, from investing in fossil fuels (Orsman 2020). In March 2018, the Dutch Green Left came first in Amsterdam's municipal elections with Femke Halsema, a former party leader, becoming mayor of the city. The Green Left, who are perhaps confusingly a member of the EGP rather than European Green Left, have introduced many radical changes to the city. Already a city of cycling, it has adopted ever more rigorous forms of recycling, for example, extracting toilet tissue from sewage and incorporating it into tarmac to reduce road noise,

> its 'auto-avoidant city' strategy will make many streets one-way and raise parking tariffs to €7.50 ($8.25) per hour. Its coal-fired power plant is shutting down, and the city plans to eliminate gas heating in homes by 2040, replacing it with electric heat pumps and centralised neighbourhood hot-water systems. A green-roof subsidy programme encourages owners to cover buildings with turf and moss. (*Economist* 2019a)

The former Rector of Edinburgh University, Peter McColl's advocacy of commons economy and free public transport is, in my opinion, an example of Greens at their best. After successive attempts, he persuaded the Scottish Greens to support free public transport. In the Scottish Parliament they were able to win free bus travel for under eighteens as part of their negotiations with the governing Scottish National Party (McColl 2020). As well as Green Party groups there are several Green Left parties. In 2017 Iceland elected a Green Left government with Prime Minster Katrín Jakobsdóttir 'budgeting $55 million over five years for reforestation, land conservation and carbon-free transport projects to slash greenhouse gas emissions' (Alderman 2019).

There is a group in the European Parliament known as Nordic Green Left (https://www.guengl.eu), which includes Green left parties as well as others on the left including Ireland's Sinn Féin.

Chris Jarvis, a former member of the Green Party of England and Wales Executive, and editor of *Bright Green*, argues that Greens can inject new ideas and create political space for action on issues such as climate change and social justice,

> Green electoral success – or at least the threat of electoral success – has driven issues and modes of thinking onto other political parties. For example, the Green Party of England and Wales was a major player in the anti-austerity movement, both on the streets and in the media. The voice of Greens in resisting Tory-Lib Dem cuts, and their subsequent receipt of over a million votes in the 2015 General Election was crucial in creating political space and shifting the Overton Window on the issue. Without doubt, this was a contributory factor in the Jeremy Corbyn's ascent to the leadership of the Labour Party.
>
> Similarly, growing interest in a universal basic income or a Green New Deal, as well as climate change being ramped up the political agenda can be seen as resulting in part from the preceding electoral threats Greens have posed. To this end, Green Parties – whether they get elected into parliaments or not – are often imperative at pushing issues onto the political agenda and shifting debates. (Chris Jarvis personal comment)

To conclude, Green Parties have the potential to raise awareness of climate change. Environmental issues were often ignored by other political parties prior to the existence of Green Parties. Greens can also introduce structural changes, from cycle paths to better building standards, to reduce carbon emissions. A minority of Greens like Peter McColl show an appreciation of a how a commons based future provides a potentially ecological and democratic alternative. However, if climate change and other environmental ills require a transformation of the mode of production, moving us from capitalist growth to commons prosperity, most Greens appear unprepared. It certainly is a more complex task than merely replacing one group of politicians with another. The strategic debate around such fundamental transformation has often been ignored by Greens and other advocates of electoral politics. Greens have integrated cultural change and non-violent direct action into their repertoire of tactics. Yet contesting elections has been the predominant tactic, and any tactic, however good, is not a strategy but only a tool that contributes to it. An alternative to Green Party politics in challenging climate change has been the construction of environmental protest groups. Extinction Rebellion and school strikes for climate have had an impact. These are examples of what academics and some activists have conceptualised as social movements. I will examine the effectiveness of such networks in Chapter Six. Greens might, rightly or wrongly, be accused of forgetting the importance of working-class agency and trade union activism. This is a topic that I discuss in the next chapter.

Chapter Five

Trade Unions for Life on Earth

A women down in Keystone (her name was Jean Battlo) once said to me: 'I'm an environmentalist, so I'm very concerned about the damage, and as a matter of fact I used to be all on that side and then I went to a meeting and heard a man say to me he was tryin' to raise his three kids and coal was his only hope.' His only hope! How much could that man have worried about climate change? (Vollmann 2019b: 145)

Back in 2009, I travelled to the Isle of Wight to support the Vestas wind turbine workers. Threatened with redundancy because of a fall in orders, the Vestas workers occupied their factory in protest. Today, I get the impression that Vestas is thriving and demand for wind turbines is strong. At the time this appeared to be a rare example of red green trade union action. The workers stayed in the factory for eighteen days to protect the 625 jobs about to be lost, Green Party members and climate campaigners supported their efforts. My friend Hugo Blanco sent a statement of solidarity from Peru. The RMT trade union acted on behalf of the Vestas workers (Blanco 2009). While I have long seen a connection between socialism and ecology, the argument that trade unions are antagonistic to green activism is often made. The shorthand being that we have to make a choice between jobs and the environment, with the former prioritised over the latter.

I first recognized a link between green politics and socialism in the early 1980s when I read André Gorz's book *Ecology as Politics*

(Gorz 1980). Having been persuaded, as a youthful Ecology Party member, that ever increasing economic growth was likely to be environmentally unsustainable, Gorz introduced me to the intimate connection between growth and capitalism. Gorz's explanation that economic growth was essential to a capitalist economy meant that capitalism was unlikely to be an ecological economic system. While my reading of Gorz provided a personal entry into left politics, for many others it provided an exit. Gorz is perhaps best known for his book *Farewell to the Working Class* (Gorz 1982). Here, he argued that the workers were no longer the key agent of social transformation, and progressive movements needed to recognise that with rapid economic change, more diverse forces would make the revolution. Gorz was not alone in suggesting that class politics had lost its relevance. Green Parties, typically, have been seen as products of post-materialism; this means that instead of a materialist politics based on economic concerns linked to class consciousness, rising prosperity in the second half of the twentieth century meant that new non-economic issues, including the environment, became more significant (Inglehart 1977). A diverse labour and trade union movement was being replaced with movements that were less concerned with economic benefit and more based on 'identity'. The post-materialist thesis of the political scientist Ronald Inglehart insists that environmental movements, specifically, were a product of a rising affluence. As society became richer with the economic growth that followed the Second World War in Western Europe and North America, we had the luxury of becoming less concerned with economic benefits and more engaged with such issues as animal welfare, civil liberties and the environment. Theorists such as the post-Marxists Laclau and Mouffe (1985), like Gorz, also argued that social movements were replacing the working class as the key agent of potential positive social transformation.

Yet when it comes to climate change, class politics in a specific sense cannot, I feel, be ignored. Climate change, as has been noted, is often seen as a result of consumption. If we consumed less, demand for fossil fuels would fall. I have argued that rising emissions and other environmental problems are products of a whole social system based on the exploitation of nature. A vital part of the system is production, and fossil fuel producers have a powerful interest in promoting the continued use of their energy sources. Workers have the potential to reengineer energy and manufacturing systems in a more ecological direction. Thus, working-class action, specifically trade union action, is vital to tackle climate change.

There is an assumption that a jobs versus environment conflict can place trade unionists on the side of the polluters. Right-wing populists like Trump and Scott Morrison have worked to gain support from working-class voters in opposing climate change action, pitching friends of the coal against friends of the Earth. If mitigating climate change means that jobs are likely to be lost, it is vital to promote green plans to create new forms of employment. There has, of course, long been a connection between environmental destruction and workplace exploitation. Green industrial action has even occurred, where workers have taken strike action to protect the environment.

Class, oppression, liberation and the environment

In 1989 the African American legal theorist Kimberlé Crenshaw introduced the concept of intersectionality. Intersectionality suggests that oppression can be multiple, based not just on class but also ethnicity, gender and other factors. An intersection is where roads cross; one is drawn to the geometry of the layout of a US city like New York. Crenshaw noted the case of DeGraffenreid versus General Motors, where black women car workers took General Motors to court for discrimination. The workers were discriminated against both as women and as African Americans. Yet the court could only recognised one possible avenue of oppression rather than act against multiple discriminations in a single legal ruling (Crenshaw 1989). Intersectionality thus challenges the notion that discrimination is, primarily, one dimensional.

Some, but not all, Marxists have criticised this notion of

intersectionality, arguing that Marx uniquely emphasised class rather than other intersections of exploitation (Bohrer 2019). Marx, it might be argued, was concerned with many aspects of injustice, however to the extent his work owed something to Hegel, he focused on how apparent disparities of power might be transformed into liberation. While elements of Marx and Engels' work dealt with gender and race, they felt that working-class agency was capable of transforming power relations. The working class of the nineteenth century were oppressed but were uniquely, Marx thought, capable of liberating the whole of humanity. They were productive, their labour power was the motor which was exploited to drive capitalism. Without the working-class, capitalism would not exist, and the creative energy of the workers was the base of a future communist society. The exploitation of capitalism gave them a motive to work for its downfall, and their existence in the factory system would tend, Marx argued, to hone a class subjectivity that would allow them eventually to overcome and transform capitalism.

In Engels', *The Origin of the State, the Family and Private Property*, the importance of gender and the existence of patriarchy was integrated into a Marxist description of social change (Engels 1909). In the twentieth century some strains of Marxism focused upon racism, imperialism and the politics of national liberation. Marxism may be read as a thing of complexity rather than a reductionist tale of single road social change. The aspect of environmental exploitation and liberation can be seen as part of the analysis too; the women suing General Motors were, for example, working in the car industry. While in the 1970s climate change had yet to become an issue, the negative environmental implications of car production were widely accepted.

The ecosocialist activist Alan Thornett was also a car worker during the 1970s. Working at the Cowley motor works near Oxford, he later became conscious that his day to day employment was environmentally destructive (Thornett 2019). Industrial production can also harm factory workers. Describing the condition of the English working class in the 1840s, in the early stages of the industrial revolution, Engels argued that poor air quality and other environmental ills destroyed lives,

> that society in England daily and hourly commits what the working-men's organs, with perfect correctness, characterise as social murder, that it has placed the workers under conditions in which they can neither retain health nor live long; that it undermines the vital force of these workers gradually, little by little, and so hurries them to the grave before their time. (Engels 1958: 109)

Working-class struggles including trade-union struggles can be seen as challenging such 'social murder'. Fossil fuel production, particularly coal production, has generated numerous episodes of social murder. In 2010, for example, 29 miners were killed in a huge explosion in Massey Corporation's Upper Big Branch coal mine, Montcoal, West Virginia. The company had failed to ventilate the mine properly, methane built up and the explosion occurred (Sweeney 2013: 13). The second volume of the William T. Vollmann's *Carbon Ideologies* sees him touring the coal fields of West Virginia, recording the devastation of both the workers and their environment (2019b). The world over, coal has been associated both with the dignity of labour and the destruction of life. The Aberfan disaster occurred in 1966, with a coal tip burying a school in South Wales, killing over a hundred children. At 9.15 a.m., 21 October, the last day before the half-term holiday, tip number seven suddenly slid down the hill. It hit and buried a farm, before completely covering the Pantglas primary school. 'The liquefied flow slide of about 100,000 tons of slurry lost energy and solidified again after hitting the school and neighbouring houses [they were] buried as completely as Pompeii.' (McLean 2009: 49)

Unlike West Virginia, and Engels' examples from Manchester, this was social murder by the state: the government-owned National Coal Board had failed to maintain safety standards.

The coal industry has become highly capital intensive in many countries, replacing workers with capital. For example, the brutal mountain top removal to source coal in the US means fewer and fewer workers are employed in mining. Yet in China and India millions of people still work as coal miners and fatalities remain high. In December 2019, the BBC reported that while safety in Chinese mines was improving due to better regulation, 333 people had died in mining accidents in 2018, and a recent disaster had claimed fourteen more lives in an explosion at Guanglong mine in Guizhou province (Anon 2019a).

Green strikes

Specific green strikes have occurred. Trade unionists refused to dump low level radioactive waste at sea, ending the practice in Britain in the 1980s (Aubrey 1991: 32). In Australia, the building workers' union leader Jack Mundy led green bans, effectively ecological industrial action. The green bans started in 1971, at Hunter's Hill in New South Wales, when the last piece of remaining open space was to be built over and turned into luxury housing. A group of residents keen to conserve Hunter's Hill, who had contacted the local council, media, mayor and other politicians without success, asked the Builders Labourers' Federation (BLF) to help them. A 600 strong meeting was held to convince the building workers to act. Responding, the BLF called a ban on building in the area and this succeed in stopping the development. Numerous other environmentally damaging construction projects were halted (Burgmann and Burgmann 1998: 176).

Trade union action on the environment reflects Marx's notion that workers have power because they produce. Trade unionists obviously have an interest in stopping what Engels described as social murder, however challenging environmentally destructive production can also challenge employment. Coal producers have campaigned against environmental movements using the jobs argument and actions such as the green ban have appeared rare. Shutting down coal, oil and gas production can lead to the destruction of working-class communities. Coal production has been an iconic marker of working-class identity in many parts of the world, and while coal production has virtually halted in Britain, the Durham Miners Gala is still a huge labour movement event.

Simply shutting down fossil fuel production and polluting industries could result in unemployment and suffering. An alternative approach has been to promote systems of just transition, planning for new, greener jobs. Such policies of transition are vital in removing social murder without killing employment. There are a number of examples of trade unions working with environmental groups to promote such plans. A Swedish car worker has argued that the expertise of car workers could be harnessed to promote an economy based on renewable energy and green public transport:

> an industry like the car industry, is not a bunch of machines and buildings. Most of all it is an organisation of people. So when humanity is facing its toughest challenge so far – to change an economy and production that has been built on fossil energy for 250 years – we need all the resources we can use to do this. [...]
>
> The car industry has an expertise in logistics, production engineering, designing for production, and quality control that could be applied to any kind of production [...] wind turbines and other equipment for renewable energy production, of trams, trains and other vehicles (Henriksson 2009).

The political economist Elinor Ostrom argued that imposing conditions and rules on others is unlikely to work and will meet justified resistance (Ostrom 1990). She suggested instead that the greater control the commoners have over the rules to conserve the commons the greater will be their commitment to respecting such rules. Likewise just transition needs to draw on the expertise of workers, the more they have ownership and control of the process, the more likely they will be committed to its implementation. One example is the story of the workers' plan at Lucas Aerospace in the 1970s. The company was in crisis, with falling demand, so the workers constructed an alternative plan to make a transition to ecologically sustainable and socially just production. Using existing skills and capital, they came up with a series of alternative products they could manufacture, instead of the planes and missiles the company was known for. The proposed projects were astonishing and ambitious, a huge array of alternative and life enhancing items, including renewable and clean energy products such as heat pumps and solar cells, together with 'a road-rail public transportation' system (Doyle 1988). Opposition from the firm scuppered the plan but it is an excellent example of how destructive production could be transformed by workers' productive creativity.

Climate jobs

In 2010 the British Trades Union Congress (TUC) voted to support a plan to create 'One Million Climate Jobs' (Trades Union Congress 2010). Retrofitting buildings to make them energy efficient, massively expanding public transport and revolutionising energy production so that it is based on renewables can all create jobs. The concept, which is similar to that of the Green New Deal, would neither solve all the dilemmas of climate change nor produce employment for all, however it would be a major positive step. By focusing on just transition it shows that climate change does not have to be at the expense of employment. Harnessing the expertise of workers is likely to make reducing carbon emissions easier. A small example I have been involved with, as a parish councillor in a rural community, Winkfield, Berkshire, was introducing a local climate change plan. The council workers, numbering no more than ten people, were asked for ideas on how the council could tackle climate change and came up with nearly fifty practical ideas for action. These were generally excellent suggestions because they built on the workers' actual practical experience. It should never be forgotten when making demands for climate change action that such action has to be implemented by workers, so workers should help shape the practices that they are asked to implement. This is both ethical and practical.

Another dimension is transforming the nature of work. Working less might be a way of reducing emissions and improving our quality of life, thus campaigns for a four-day working week can be seen as an ecological measure (Rahm 2014: 178).

New trade unionism and social movements

There are a number of factors that limit the effectiveness of trade union action on climate change. Anti-union legislation means that green strikes are likely to be illegal in some countries such as Britain. Trade unions have generally been becoming weaker in recent decades, with membership declining steeply in many parts of the world. The radicalism of trade unions may also be limited; trade unions might help workers to gain higher pay and better workplace conditions, but workers may pursue relatively minor reforms rather than embracing systematic and fundamental change. In both Western Europe and the USA, rapid deindustrialisation over recent decades has contributed to a collapse of unionism in traditional areas of employment. Service sectors are far less unionised and an apparent neoliberal turn in contemporary capitalism has allowed firms to outsource jobs in many areas of the economy to supposedly self-employed workers. Such self-employment is part of the gig economy, so called because workers turn up to gigs when needed, rather than working consistent hours. This neoliberal turn often involves a transition from secure employment, with pension and sickness rights, to potential poverty. Thus, while trade union consciousness of climate change and other ecological issues might have risen sharply, trade union capacity to act seems to have diminished.

One encouraging trend has been the growth of militant, often new, grassroots trade unions. In Britain, workers in the gig economy have organised to improve their conditions. I am not aware of a specifically ecological element to this new unionism, but it has helped cycle couriers and others who deliver using bicycles to bargain more effectively:

> In September alone, Deliveroo workers have struck in at least 16 cities/zones [...] in response to ongoing relentless attempts to force down wages. [...] strikers have used tactics like large flying pickets and blockades of dark kitchens (delivery-only kitchens, often based in industrial sites) in order to disrupt Deliveroo's service. Many strikes, such as the one that took place in Brentford on the evening of Friday 27th, are so successful that they have forced their bosses to turn off the app completely. (Cant 2019)

We can conclude that working-class activism is an essential part of the struggle against climate crisis. Workers have the means to change society in a greener direction, workers in a variety of industries have production expertise and can thus potentially plan greener forms of manufacturing. While there is a possible contradiction in that climate change action can lead to job losses, in many areas such as fossil fuel production employment levels are already declining because of automation. Nonetheless, if workers are to have a stake in transforming production in a greener direction, serious attempts to create just transition plans are vital. Vollmann's example at the head of this chapter is significant; while recognising the environmental ill effects of coal production, workers may feel that they have no alternative means of feeding their family. Ultimately, transforming how we produce is an important part of changing the current destructive mode of production.

The likely success of green trade unionism depends on effective tactics to win arguments, mobilise workers, increase solidarity and win struggles. This chapter started by introducing Andre Gorz's suggestion that we need to say farewell to the working class. The farewell might assume that social movements are different to trade unions and represent a more effective agent of social change. My assumption is that to some extent, rather than being entirely unique and separate entities, political parties, trade unions and social movements overlap. The climate strike movement started with school students walking out on regular Fridays to campaign for a future. Its advocates have called, with some success, for adult workers to walk out and join them. The new militant grass roots trade unions, organising the unorganised, particularly in the service sector, may resemble social movements. I would go further and suggest that when we study social movements, we study not just distinct organisations and networks, but particular tactics and methods. The right has been successful to some extent in mobilising to make climate change worse, a theme taken up directly in a later chapter. To defeat the right and to work for social change, including trade union action on climate, we would do well to think more deeply about social movement tactics. There is now a large academic literature that studies social movements, investigating the how and why of their successes and failures. Examining this is part of the process of creating the change we need to face the challenge of climate crisis. This is the subject of Chapter Six.

Chapter Six

Extinction Rebellion

Firstly, how do we have sustained social movement agitation that is constantly chivvying the state and business, forcing them to make promises and also watchdogging them relentlessly into keeping the promises? How are those social movements able to sustain themselves, without being co-opted and/or repressed? How can social movements avoid the smugosphere, the emotathons and the theme park of radical action? (Hudson 2018)

Extinction Rebellion (XR) was established in Britain in April 2018 (Hallam 2019: 20). Members of XR believe that climate change threatens extinction for many species including, perhaps, human beings. They take disruptive but non-violent direct action to raise the demand that a climate and ecological emergency is declared and that serious steps are taken to move towards zero emissions. They have spread globally. In Britain they demand that emissions must be net zero by 2025 with climate policies implemented after they have been formulated by a citizens' assembly. Founded after an appeal by a Swedish teenager, Greta Thunberg, in 2018, Fridays for Future involves students leaving their classes and taking to the streets to demonstrate. Participants argue that while education is important they will not have a future if climate change wreaks havoc, so walking out of school is more than justified.

Both XR and the Fridays for Future climate strikes can be described as social movement organisations. Academics see social movements as more militant, more disruptive and less formal than pressure groups. For example, pressure groups, like Greenpeace, have a formalised institutional existence and work like businesses with employees, etc. Likewise, social movements can be contrasted with political parties because they generally don't contest elections and seek formal political power in parliaments or senates. Of course, climate change social movements can overlap with both pressure groups and Green Parties.

While XR and Fridays for Future are perhaps the best known climate change protest movements in recent years, they are not alone. Local community campaigns involving demonstrations and direct action around climate issues have occurred both in Britain and internationally. One example is anti-fracking protests, which have combined local concerns for protection from pollution and even earthquakes with campaigning against climate change. Another aspect of social movement campaigning has been disinvestment, which aims to get institutional investors such as pension funds to remove monies invested in fossil fuels.

Back in the late 1990s I completed my PhD on the movement against motorway construction that was prominent in Britain at the time. Members of a radical ecology group, Earth First! (EF!), kick started protests that attempted to disrupt the building of new motorways (Wall 1999). While they failed to directly stop the projects that they had disrupted, the movement successfully raised awareness of the environmental ill effects of road building and helped halt additional construction. A host of other environmental targets were challenged, and climate change concern was a feature of this mobilisation. I wrote a hundred thousand words, attempting to understand the factors that made this, in my opinion, a successful social movement. While I haven't a hundred thousand words to use here, I think it is worth looking at some of the reasons social movement researchers have argued that mobilisations succeed or fail. 'Success' or 'failure' are contested terms, contexts vary as well but there is a vast literature on social movements, and some of it might be useful to activists in networks such as XR or the school strikes.

Extinction Rebellion

XR came to attention with a protest of over a thousand people in Parliament Square, London in October 2018 (Taylor 2018). In the following weeks they graffitied government buildings, blocked main roads in the capital and occupied five bridges across the Thames. This gave them publicity and they expanded quickly, recruiting new activists to become a sizeable movement. XR, from modest beginnings, grew and grew. In April 2019 XR occupied sites in central London including Piccadilly Circus, Oxford Circus, Marble Arch, Waterloo Bridge and Parliament Square. Areas were taken and a party atmosphere ensued. Numerous arrests were made. XR has grown very fast, has had a huge media impact and perhaps more than any other mobilisation has attracted huge attention to climate change. The XR model is apparently derived from momentum training, a US system of promoting direct action. The writer Graham Jones has described this as involving a clear structure which is presented in an accessible way to would-be supporters and passed on via mass training. He further notes that the direct action aims explicitly to polarise the public. These working principles might be thought to be,

> the *DNA* of an organisation, its *replication*, and a *catalyst* for kicking off that process. The aim is to enter into a growing feedback loop, whereby the direct action attracts attention, creates an inflow of new people into training sessions, who enable the organisation of more and larger autonomous groups, which can then do larger actions etc. Ultimately enough people join the moment to allow for actions which can shut down urban centres for extended periods of time, just as you would have in, say, a general strike. (Jones 2019)

The XR method is to be disruptive, maximise arrests, clog the courts and push the government into serious action. While

XR raised awareness of an emerging climate crisis, it has been criticised. There have been a number of thoughtful discussions analysing a variety of aspects of the strengths and weaknesses of XR. Chris Saltmarsh, who works with the student-based People and Planet, noted, 'the willingness of so many to take radical action is refreshing and inspiring' but criticised XR's analysis,

> Framing climate breakdown as a moral issue rather than as political, Extinction Rebellion seek to universalise their movement. But they ignore the power relations which structure the crisis. If politics is about who uses power, when and for what ends, the story of climate change is a deeply political one where a political and economic minority inflict the injustices of climate breakdown on a systematically disempowered and dispossessed global majority. (Saltmarsh 2018)

XR has been criticised for showing little awareness of the influence of racism and other forms of oppression in the context of climate change campaigning. A XR tactic, of chanting 'we love the police', has not been met with universal acclaim by other climate activists (Punk Academic 2019). Concern has also been expressed about the quality of advice given to those arrested and lack of support during the process of being held in cells and going through the courts. Green and Black Cross withdrew their legal support work for XR after criticising XR's approach to protecting participants during direct action (Green and Black Cross 2019).

XR continued to grow fast during 2019 and has had much success in mobilising support for climate change action in Britain. Maintaining movement growth may be challenging. It would be easy for a combination of bad publicity, arrests and exhaustion to lead to collapse. Disruption of public transport might enrage potential supporters. Seeking to disrupt London's underground network, two XR members climbed on to a tube carriage. They were dragged off and attacked by enraged morning commuters, trying to get to work.

There were chaotic and shocking scenes as one activist was kicked and stamped on by the furious mob after he tumbled off the roof of a Jubilee line train at Canning Town station in east London.

Two climate change protesters had halted service and unfurled a banner – reading 'Business as usual = death' – but they were jeered, and one was pelted with food and other objects by commuters who didn't want to be late for work. The banner was ripped out of one activist's hands and the other was dragged onto the crowded platform and 'kicked in' (Kitching *et al* 2019).

While XR activists might argue that little could be more disruptive than extreme climate change, and the transport disruption was perhaps a product of a few individuals in a decentralised movement, such dramatic direct action may lead to lost sympathy.

Social movements

While there is an extensive research literature on protest movements it is impossible to determine precisely which particular tactics and approaches are most likely to be successful. Researchers disagree, and what works in one context may not work in another context. However, social movement research can usefully encourage those involved in protest to see their actions as 'alien' rather than inevitable. What I mean by this is that a Brechtian perspective may help us to appreciate what we are doing and why. Bertolt Brecht, the German revolutionary playwright, used the term '*Verfremdungseffekt*', variously translated as 'alienation', 'distancing' or 'estrangement', to explain his method. He wrote plays that 'distanced' the audience, making them aware of how a play was put together to make it easier for them to understand techniques that might be used to manipulate audiences. Rather than using a battery of techniques to make a play look 'natural', so the audience would forget, at least for a moment, that it was constructed, Brecht sought to show that it was artificially assembled (Jameson 1998). The aim of '*Verfremdungseffekt* is to puncture the complacent acceptance of either character, motive, narrative, incident or resolution, as "fixed" and "unchanging" or "obvious" and "inevitable".' (Brooker 1988: 63) We need to do this with social movements, showing that what we view as a common sense methods for social change can be rethought. Being conscious that things can be done differently means we can be more conscious of what we are doing, and perhaps control it a little more.

XR activists and School Strikers are driven by the urgent need to halt climate change. It threatens our future, so we protest. Social movement theorists argue, broadly, that such protest movements can be described and analysed. They are not simply spontaneous uprising, driven by particular concerns such as anti-racism, equality or climate change. Instead they have to be consciously constructed and are products of social circumstance.

Social movement theory has, like any academic study, moved between different dominant perspectives. In the 1950s and before it was largely a product of functionalist sociology. Functionalists, as the name implies, argue that societies tend be functional. Functionalist sociologists such as Talcott Parsons argued that apparently unusual and perhaps troubling aspects of social life tend to conserve and maintain social order. Social movements from this perspective were evidence of contradictions and potential problems in society, they could make such problems visible so that they could be addressed by politicians and police. Protest also allowed those with grievances to act and let off steam. Protest from this perspective, paradoxically, prevented revolutionary change and rupture (Wall 1999).

The emergence of the US civil rights movement, including the Freedom Summer campaign, in the 1960s led to the growth of Resource Mobilisation Theory (McAdam 1986). This provided a challenge to the functionalist approach, focusing on the rationality of social movement activists. In a way it was an economic theory of protest. According to this view we are motivated by broad self-interest, protesting because it will bring us benefits and protest succeeds where activists can mobilise resources most effectively. Resources might include money to fund campaigns, donations of tents for occupations, carpenters to build protest boats, etc, etc. It has perhaps some analytical value, yet personal self-interest is unlikely to fully explain the motivations of XR climate rebels. Neither are resources the only factors explaining relative movement success or failure. However it does take activism seriously rather than seeing it as an outgrowth of impersonal social forces and resource consideration do help movements. Certainly, Fridays for the Future might be seen as rational response of school students trying to defend their future. In turn movements like XR have been impressive at gathering resources, which has helped them to grow rapidly.

The more US influenced Resource Mobilisation Theory approach can be contrasted with the broadly European research perspective that emerged in the 1970s and 1980s. The European perspective, to simplify, introduced a cultural dimension based on beliefs and practices (Martin 2015: 93). Even if we are the pleasure seeking rational maximisers described by some economists, what brings us pleasure is not a product of the mirror of unmediated biological drives. What we desire so passionately is produced partly, at least, by cultural factors. This suggests a more complex understanding of human motivation than that of rational choice and resource gathering. Culture is incorporated into the practices of social movements including those fighting the climate crisis. XR actions have a vibrant dimension of music, sound and art, which draws in supporters and increases visibility:

> The Extinction Rebellion protests, which are set to conclude today (25 April) with an occupation of the financial district, have included a stage for musicians and artists since they began on 15 April. Draped with a sign that says 'This is an Emergency', the stage in Oxford Street has featured Massive Attack, Alabama 3 and Aurora Dawn, Caligula, Inner

Terrestrials, The Undercover Hippy, Mazaika and more
There has also been impromptu folk music sessions, classical music performances and a 'Play Piano for the Planet' event. (Anon 2019b)

I am reminded by Marc Hudson, a social movement scholar who was kind enough to read an earlier draft of this chapter, that 'no culture is entirely inclusive. What works to bond some, repels others ... we mustn't romanticise music, etc as inherently progressive, either' (Hudson personal comment).

Another contribution of the European approach to social movements was to suggest that macro factors of broad social and political change also helped explain cycles of social movement mobilisation. A variety of social movement theorists argue that social movements move through cycles of birth, growth and decay. They rarely move in a smooth upward direction. Whatever we think of the varied academic perspectives, this description indicates that activists might benefit, perhaps, from understanding the likely ebb and flow of their movements.

European perspectives also helped give rise to an understanding of social movements as specifically 'New Social Movements'. Yet the description of new social movements as movements of identity contrasted with old social movements based on more prosaic economic concerns is at best an over simplification. Climate change too has material effects; it is far from being a post-materialist luxury, for example, increasing extreme weather events such as storms, floods and droughts make life more challenging. The increasingly common fires in parts of the world such as Australia and California, destroy people's homes and on occasions lead to loss of life.

Social movement theory has proliferated, with new branches and concepts. For example, the study of social movements and emotion has provided one new dimension (Jasper 2011). To proceed I will introduce some concepts used by researchers to help understand social movement. Broadly these descriptive categories reflect a fusion, which has become central in academic debates, between the US based Resource Mobilisation Theory and the more cultural and political understanding of the European theorists.

Political opportunity structures

Political opportunity structure approaches suggest that social movement mobilisation is a product of political processes. Social movement academics like Sidney Tarrow in his *Power in Movement* (Tarrow 1998) argue that movements work within a political context and this helps explain their potential success. He suggested that if the political system becomes more open and if opponents suffer from division, movements are more likely to grow. In the context of the 1990s, researching EF! and the anti-roads movement, I stressed the closure of parliamentary political opportunities in Britain. The electoral system made it difficult for the Green Party to win elections, the existing Conservative government, and the opposition Labour Party, were unsympathetic, at the time, to environmental demands, so direct action protest became a more attractive option (Wall 1999). However the political opportunity structure did not directly produce the protest. Social movement activists, theorists have sometimes described them as social movement entrepreneurs, have had to build protest movements. The political context makes this easier or more difficult but does not automatically lead to the creation of a movement.

XR founders Roger Hallam and Gail Bradbrook deliberately sought to build a protest movement rather than writing to their MPs, because of their assessment of potential political opportunities in Britain. One weakness of a political opportunity structure's approach is that it might become so extensive that it becomes vague. A whole series of factors might be defined structurally as political opportunities, from electoral systems to media culture and changes within and between political parties. In turn, such factors may create contradictory opportunities and effects. The skilled political entrepreneur might seek to understand different forms of opportunity and how they might interact. Perceptions of political opportunities may be as important as political opportunities; we imagine that something is politically possible, so we organise to achieve it. The organisation opens up the system and the illusion of opportunity convinces us to organise. Opportunities might be so complex that an element of change emerges. Perhaps movement entrepreneurs throw the dice but in doing so they consciously, or unconsciously, calculate the odds.

Action, direct, indirect and effective

A particularly useful area for making the familiar unfamiliar so that we can be more conscious of what we do and why, is movement tactics. It seems natural perhaps, if we have a grievance, that we take a particular form of action, writing to a Member of Parliament, signing a petition, etc. For social movements particular forms of action may appear obvious. The Fridays for Future movement goes on strike, withdrawing participation in school, and marches. XR takes part in disruptive direct action, typically blocking roads but sometimes even seeking to disrupt rail and underground trains.

Social movement theorists describe such action tactics as 'repertoires of contention' (Tarrow 1998: 23). Particular repertoires, such as the barricade, have been used over hundreds of years. Researchers have examined sources such as police records and newspaper reports to map the occurrence of such repertoires. Repertoires may be surprisingly stable, while there is vigorous debate around the concept of 'new social movements' repertoires such as the strike might be traced back as far as ancient Egypt (Wilkinson 2013: 358). Social movement researchers argue that particular repertoires are learnt and transmitted; they are familiar both to activists and the authorities who are being challenged. My feeling is that innovating new repertoires or variations on repertoires may give protesters an element of surprise allowing them to be more effective. The study of repertoires has in the past suggested a theatrical ritual basis with both activists and authorities moving through scripts. Rewriting scripts is the basis of something new, but if we fail to recognise the scripted nature of our action, constructing new narratives may be difficult.

Spare time

Movements are born, ebb and flow, grow, but eventually die or at least diminish. When individuals disengage and stop being active, movements become weaker. While there are all sorts of large-scale influences from changes in the economy, social media and political opportunities that effect how active individuals are, how they join and leave movements, social movement theorists have also used the simple concept of 'biographical availability' to help explain participation. Doug McAdam (1986) who studied Freedom Summer, argued that the simple fact of having spare time so that individuals could be available to take part in the voter drive was important. It is easy to think that because climate change is such an urgent and dramatic issue that individuals will join movements such as XR. One lesson of social movement research across varied perspectives is that an issue, however important, does not automatically generate a movement. Movements have to be made, to be constructed. Even where we feel passion we may have commitments that make it impossible for us to easily participate. Biographical availability suggests that at some points in an individual's life cycle they may have more time to protest. An intersectional understanding is perhaps missing from some of the classic academic literature; those with multiple oppressions are likely to be under more pressure and have less availability.

While it is a truism that some groups, like pensioners, may have more time to protest, the notion of availability should be taken seriously by those trying to sustain climate and other social movements. For XR, an emphasis on maximising arrests might be counterproductive. Clogging up the legal system puts pressure on the political process to gain change and helps draw attention to the climate emergency. However it makes it more difficult for many individuals with caring or work commitments to take part. Experience of the legal process of being arrested, held in cells and moved through the courts can be challenging.

Militancy may lead to results but it has a cost in terms of potential participation. This needs to be assessed if movements are to be sustained for long periods of time. Marc Hudson describes the practical importance of biographical availability with his usual bluntness and clarity:

> Do you have kids? Sick aging parents? A (horrible) long-hours job? Other responsibilities that mean you don't have the time, energy or inclination to sit in boring meetings that never seem to achieve anything? That's biographical (un)availability, and social movements are mostly hopeless at coping with it. (Hudson 2017)

In turn, movements might want to consider how they continue to exist in periods of relative quiet and how they build, rather than diminish, activist capacity. A focus of criticism of Extinction Rebellion from Hudson, is that movements are difficult to sustain; initial enthusiasm can disappear. He implies that movements should be conscious of how to conserve and sustain activism, without attention to this aspect, mobilisation can be short. While he makes no claim that his analysis of emotional cycles in social movements is universal, basing his analysis on social movement mobilisations in Britain over the last 25 years, my feeling is that it has wider application. He identifies the emotacycle, a cycle of mobilisation, kicked off by rising enthusiasm that is eventually used up, leading to demobilisation,

> a sequence of relatively predictable events and decisions by social movement actors (individual and group) to engage in ritualistic and familiar actions (marches, protests, camps) that allow for the performance/release/management of certain kinds of emotions, regardless of whether they ultimately achieve the stated goals of the individuals and groups. (Hudson 2019)

XR lacks an apparent focus on movement energy conservation; initial enthusiasm can dissipate. The question of how we move from emotionally charged, but perhaps short lived, cycles of protest to more enduring activism is not easily answered. However, I feel the concept of base-building, to create enduring communities of resistance, through long term community activism is one possible answer to this question. Base-building is a topic explored in Chapter Ten.

Individual and collective identities

Resource mobilisation theorists argue that movements depend on gathering resources and perhaps the most important resource for action is the activist. Biographical availability is a precondition for intense movement involvement such as occupying buildings or blocking roads but being available doesn't explain involvement. Just having spare time will not mean we will become active in XR or similar bodies. In recent decades social theorists have increasingly emphasised that our personal identities, rather than being 'given', are constructed: we are shaped by a number of factors and equally we shape the societies and forces which affect us. Social movements need two forms of identity, our individual identity as an activist, and the collective identity of an entire movement. Both are necessary and both are multiple, with different personal and collective identities overlapping and even in potential contradiction. We might see ourselves as a teacher and a parent, we might belong to an anti-fascist movement, worship at a mosque, identify as vegan. Activist identities are vitally important for movements, for example, while we might be concerned with climate change and have available time, we might not gain an activist identity and participate in strong collective action such as a climate strike. XR has attempted to extend their movement by creating XR community groups based on ethnicity, profession or age. These range from police and architects, to grandparents, to Sikhs and permaculturalists,

XR Community Groups are a way for rebels to connect and work together through communities of shared self identity rather than of shared location. (For example, faith, profession, ethnicity and sexual identity.) (Extinction Rebellion 2020)

Theorists have tried to understand processes of identity construction. Networks are important, if we have friends, family members and other associates involved in a movement, we are more likely to take part and embrace an activist identity. Increasingly, identity construction in movements has been seen as strongly influenced by language. Movement entrepreneurs must construct their messages carefully to gain support. This points to the fact that social movement theory is a manifestation of wider forms of social theory. If all identities are constructed rather than being 'natural' the processes which help create us are worthy of study. I am reminded of the words of a Pet Shop Boys' song 'I never dreamt that I would get to be the creature that I always meant to be' (Sinfield 1998: 82). This refrain, sang by Neil Tennant in 'Being Boring' is about the construction of gay identities but might be discussed in reference to identity construction in general. Perhaps we can build identity more consciously (Einwohner and Reger 2008)?

Framing climate

A significant problem in raising awareness about the seriousness of climate change and the need for strong action is communication. Climate change is scientifically quite complex, solutions are contested, and nuanced understanding is necessary. Political communication perhaps involves identifying simple causes and providing clear solutions. To gain attention it may be necessary to paint one's proposals in bright and lurid colours. The opponents of climate change action have often proved more sophisticated in challenging the necessity of action than its advocates.

Frame analysis has been used to examine how messages are constructed by social movement organisers. The term 'frame' is derived from the work of the sociologist Erving Goffman who suggested that meaning is constructed by emphasising some aspects of reality and reducing the significance of others (Goffman 1975). We frame an argument as we would frame a picture; the frame also enables us to interpret what is going on. The analysis of 'quick and slow thinking' suggests that given the complexity of reality, we often use preformed assumptions, stereotypes, perhaps to understand quickly what is going on (Kahneman 2012). Human consciousness does not act as a mirror which reflects reality but interprets reality on the basis of previous assumptions. Logical argument is often less important than an emotionally charged image in shaping our commitments; a supposition explored in Chapter Eight, which examines the communication of climate change science.

Social movements, it has been suggested, must go through a three-stage process of framing if they are to successfully mobilise support (Snow *et al* 1986). A problem is identified, a cause is outlined and a course of action is suggested. For example, climate change movements identify climate change as a threat to human survival, a cause is outlined in terms of continued commitment to fossil fuels and action recommended such as declaring an ecological and climate emergency.

Frame analysis has developed beyond describing the three basic tasks necessary to mobilise others. Frame bridging connects different frames, and is useful for building coalitions. XR, for example, might use anti-capitalist framing to bridge to those on the left, and as noted, have created diverse XR community groups to widen their potential activist base. Frame amplification means heightening the importance of the elements of the frame. XR is particularly effective at this. Climate change is framed as threatening our collective existence, and what could be more important than that? Frame extension means incorporating new concerns, so as to draw in new supporters. Frame transformation is the perhaps challenging task of changing our basic system of interpretation, taking us, for example, from a liberal democrat understanding of social reality to one of anti-capitalism or deep ecology.

Framing can be used to gain media attention, which in turn is used to spread a movement's message further. Social media has allowed political movements to communicate without direct access to traditional forms of monopolistic media such as television or newspapers. However social media platforms are also becoming increasingly corporate. Rather than promoting social change, the likes of Facebook will support the powers that be. From US attempts to access data, to the great firewall of China to the suppression of social media accounts supporting Kurdish liberation, what is open becomes closed. Climate change movements must keep innovating if they are to communicate effectively.

The strength of weak networks

Another concept in the social movement studies toolkit is the notion of networks. While biographical availability is necessary to participation in a movement and framing can persuade individuals to become active, network links are a strong predictor of involvement. Simply put, we are most likely to join a movement if we know other people who are already involved. One way a movement can grow quickly is to make links with pre-existing networks. In a highly cited paper the sociologist Mark Granovetter noted 'The Strength of Weak Ties' (1973). What he means by this is that a movement that makes weak ties to many other movements can communicate its message to large numbers of people. The contradiction which is worth exploring is that framing can be used to induce strong commitment from activists, but strong ties may make it more difficult to communicate persuasively to large and diverse numbers of people. The 'new communist movement' in the USA and Trotskyist parties in Britain may have enjoyed intense political involvement from relatively small numbers of committed 'cadres' at the expense of involving larger numbers of individuals (Elbaum 2018; Kelly 2018). Sticking with Marxist militants as an example, Gregor Benton, in *Mountain Fires* a study of Maoists in Southern China in the 1930s, shows how they created wider support by building network links with

Daoist fighting bands, kinship groups, varied ethnic minorities, etc (Benton 1992). For any organisation or movement, from a conventional political party to a social movement or a revolutionary group, individual commitment requires intense dedication but weaker ties to diverse networks aid recruitment and communication.

EF! was established in Britain, initially, as a deep ecology group. EF! in the US had faced controversy because of the Malthusian and right-wing beliefs of some of its founders. Indeed this was the subject of criticism by Murray Bookchin (Bookchin and Foreman 1991). In Britain they grew fast as a loose network based on using disruptive but non-violent direct action primarily to oppose road building but also to challenge a number of other forms of environmental damage. EF! networked with existing green organisations to build a base of a couple of hundred activists. Their momentum increased by drawing in a diverse array of groups from animal liberationists to local conservation campaigns, anti-war networks to trade unionists. Deep ecology was largely replaced by more social framing of environmental issues. While EF! still exists,* like many movements they declined; social movements typically move through a wave of growth followed by a falling-off of activity (Wall 1999). Their initial growth does appear to be another illustration of the value of weak ties to numerous networks, even for groups that require intense activist participation in direct action.

Power, strategy and change models

Social movements and indeed political parties have a performative aspect based on constructing and celebrating particular identities. From a Young Conservative garden party in 1950s England to a group of banjolele players at San Francisco Occupy, part of political and social commitment is about performances and practices to mark who we think we are. However while identity can never, as discussed, be entirely separated from politics, movements along with parties are also about power. Social movements have stated objectives they wish

* https://www.earthfirst.org.uk

to achieve, to achieve them they have strategies and use tactics. How they construct and implement strategy is shaped, at least partly, by their assumptions about power.

Power is a highly contested concept with numerous definitions and debates. Put most crudely, it might be our ability to get what we want. Yet what we want might be something that we lack control over. If power is about gaining the object of our desire, might a more powerful force determine or at least influence the nature of our desire? Power might have a macro element with relatively enduring forces of social class and other sociological and economic factors having an influence. The French theorist Michel Foucault has suggested, in contrast, that power works at a micro level, a product of small-scale interactions, and is everywhere. Power may be theorised as either power over, which means the oppressive power over others to get them to do what we want, or as power with, productive power, that constructs rather than oppresses (Hay 2002: 170).

The encyclopaedia of power would have many volumes. Consciously or unconsciously, diverse conceptions of power have an effect on what social movements do. Much social movement activity might be based on a kind of militant liberal democratic assumption. This might suggest that we live in societies which are pluralist, with diverse interest groups, and protest can be used to influence policy making decisions. The aim of a protest is to make an issue more visible, to gain increased public support by protesting, and by using such support to make politicians take particular decisions such as banning tar sands exploitation or subsidising solar energy.

I think this, to some extent, broadly informs the approach of groups like XR and Fridays for Future. Power, from this perspective, is largely a product of decisions made by politicians in parliaments and senates. Public opinion shapes what politicians do, so by influencing public opinion change can be achieved. This fits in with a strategy of bearing witness or even shaming. Vibrant action can be used to force an issue onto the agenda and to gain policy change. Roger Hallam, one of the XR founders, went on hunger strike successfully to force King's College London, where he was studying for a PhD in 2017, to divest from fossil fuels (Hallam 2019: 46). Scaled up to a national and then international campaign, pressure is brought to bear on politicians to call a climate emergency. With a broad pluralist frame, it is acknowledged that corporate interests have a degree of power to promote their demands to keep extracting fossil fuels, so protest raises the power of those who seek to challenge climate change. The approach of EF! back in the 1990s was different. Some of those involved with the anti-roads protest, in fact many, sought to change policy. Non-violent direct action in an age before Facebook and Twitter was based on gaining mass media attention. Dramatic action gained news coverage and news coverage gained the attention of voters and policy makers. However EF! largely rejected mediation by the media as a way of influencing the mainstream policy making process. In the USA, EF! was motivated by a desire to sabotage the process of environmental destruction. Inspired by the novel *The Monkey Wrench Gang* (Abbey 1975), they published a handbook of ecologically motivated sabotage, which included advice from cutting down billboards to how to disable vehicles by putting sugar in their petrol tanks (Foreman and Haywood 1993). A cynical attitude to established policy making inspired acts to slow down the forces of destruction.

EF! in Britain, inspired by anarchist, animal rights and anti-war approaches, including the women's peace camp against cruise missiles at Greenham Common, stressed collective protest rather than individual or small group sabotage (Wall 1999). Direct action, in the words of the anti-motorway network Reclaim the Streets (RTS), was 'the preferred way of doing things' rather than a desperate last resort (Wall 1999: 192). Direct action was directly disruptive, war if you like by non-violent means, that would make it too expensive to build motorways, extend airports or engage in other environmentally destructive civil engineering projects. The tactics of occupying sites that were due to be demolished was used to slow such projects and to build alternative communities. In contrast to XR the aim was to entirely by-pass the state. The assumption was that the state was never a force for good but worked within a capitalist system to do the will of the rich and powerful including the polluting corporations. Anarchist anti-capitalist assumptions about power and change were combined with creative tactics of disruption. The intention was to act directly, not to use direct action as an appeal to the public, or to act as a way of influencing politicians.

Direct action involves sacrifice, but occupations and street parties, promoted by EF! and RTS, were often joyous affairs attracting thousands of participants. Although these forms of mobilisation peaked in the mid to late 1990s and fell away, they influenced new generations of activists. Permanent protest camps at fracking sites draws on forms of protest used by anti-motorway campaigners like EF! in the 1990s and women peace campaigners at Greenham Common in the 1980s (Wall 1999).

Elements of XR's activism resemble the EF! approach in that space is occupied, often to be used joyously and creatively. It might be added that whether consciously or not XR works with a notion of urgency that might be seen as a little moralistic. The climate crisis is a crisis of extinction, so we should all react if not with panic at least with the resolve of a citizen who arrives at their door to find their home is in flames. EF! aimed to disrupt the forces of destruction, making it difficult and expensive to build motorways, XR seems at times to be keen to disrupt life. By disrupting our routine, we are moved to join the rebellion. Or that is the theory.

There is, as noted already, a potential contradiction that in promoting passionate action on climate change, bridges to a larger population are broken. The issue of rising levels of denial around climate change in countries such as the USA and Australia needs addressing. In Australia, for example, militant action on climate change has failed to win over a clear and substantial majority of the population. The result being the election and re-election of governments keen to mine more coal despite rising damage from droughts and forest fires. Some attention to frame bridging is necessary. If 3.5 per cent of the population is actively engaged in climate action, XR's narrative suggests, policy makers will fall into line. But if another 3.5 per cent are engaged in counter movements, it may be that nothing moves. In fact the 3.5 per cent figure has been subject to detailed critique (Ahmed 2019; Berglund and Schmidt 2020).

The research of Chenoweth and Stephan (2011) in *Why Civil Resistance Works?* that Roger Hallam drew upon was based largely on movements aimed at overthrowing governments in case studies including Iran, Palestine and the Philippines. Hallam (2019) hints that we have a political regime hostile to climate action that must be overthrown and replaced.

> Let's be frank about what catastrophe actually means in this context. We are looking at the slow and agonising suffering and death of billions of people [...] This is what our genocidal governments around the world are willingly allowing to happen [...] governments knowing the impact of climate change but continuing to support the fossil fuel industry, the result will be the destruction of many nations, species and cultures. There is no greater crime. (Hallam 2019: 16)

I am not sure how useful this. As noted throughout, my assumption is that we have an economic and social system with features that tend to promote environmental degradation. Transforming such a system, assuming I am correct, is a little different from removing a particular government or even many governments. Assuming I am wrong, there are plenty of other models of the task at hand that might plausibly describe a different understanding than is suggested by the academic model of social movement activity that Hallam draws upon. Either way case studies of non-violent uprising around the world may not be fully relevant to understanding social movements challenging the climate crisis.

Social movement engagement might work to build bridges

by engaging with a variety of communities; incidentally this a major part of Hallam's analysis. As noted previously Extinction Rebellion's attempts to build bridges with the police have proved controversial. Hallam argues explicitly that a moral debate over whether the police are 'good or bad' should have no bearing on movement strategy (Hallam 2019: 32). In revolutionary situations if members of the military can be won over, the revolution will win. This is one of the reasons why Chenoweth and Stephan suggest non-violence is more effective than violence. Hallam's analysis draws upon the political opportunity assumptions discussed earlier in this chapter. Like Sidney Tarrow, he argues, that divisions in an elite, provide an opportunity for movement success. Thus open and respectful interactions can be seen as vital to win over the police and others who function to protect, in Hallam's view, a genocidal state. Hallam's analysis is strong in that it attempts to focus on material strategic questions, putting forward his understanding of the most effective way forward in challenging the climate crisis. I am not convinced that it is entirely accurate, certainly looking at the role of the police it is open to question.

EF! mobilisation in the 1990s may have been weakened by police infiltration. Police might be appealed to as individuals with a concern for climate, but it is worth noting that protest policing has a long history and is sophisticated. Friendly interaction with protesters is a way of collecting intelligence that can be used to reduce protest or redirect it. Police infiltration involved many examples of undercover intelligence operators becoming trusted members of environmental protest communities. The spy cops scandal occurred because there were cases of police going undercover and fathering children with women activists (Lewis and Evans 2013). It might be argued that it is good to turn enemies into allies but letting enemies enter your movement is a sure road to defeat. Hallam does not as far as I know address this issue of police infiltration, beyond suggesting that if movements are open and large enough, such infiltration will be ineffective.

Much of Hallam's analysis and the practice of XR advocates,

as noted, makes a strident appeal to avert catastrophe and aspires to mobilise a percentage of the population to take action. Dealing with climate change is likely to be challenging to many of us who drive cars, enjoy taking flights abroad or whose employment depends on ecologically dubious activities. There is a danger that the XR rhetoric of emergency can be used to dismiss these varied concerns. Workers in particular need to be reassured that they will not become unemployed as a result of climate change policies. Detailed conversion plans to substitute jobs in fossil fuels for alternative forms of production need to be constructed, and to be constructed with the involvement of workers. While, as noted, lifestyle change may have to be part of action to prevent or slow an acceleration of temperatures towards hot house earth, it is important to focus on how this can be made as practical and pleasant as possible.

Social movements: defensive or revolutionary?

Social movements at first glance look like rather defensive institutions concerned with limited policy reforms. On occasions they might spark civil disobedience that might lead to regime change. At their most radical they have the potential to change codes, so as to challenge prevailing beliefs and practices in a fundamental way (Melucci 1996). In doing so they can contribute, perhaps, to the radical transformations needed to deal with climate change and other ecological ills. Culture is part of a mode of production, or if you prefer, an alternative terminology, a social formation or society. Social movement mobilisations have a number of other implications. We tend to think of heroes and villains, so we have noble social movements trying to end slavery, stop war or protect badgers, and evil forces protecting slavery, promoting war and lobbying to kill brock. In reality, the world is less binary. Nonetheless as well as environmental social movements, we also have anti-environmental counter movements, which work to prevent serious action on climate change. As such they should be studied and understood, condemning or ignoring them is inappropriate. Social movement theory is perhaps artificial or even arbitrary, political opportunity structures, framing, and other conceptual tools, are perhaps useful in studying a range of collective political actors from revolutionary groups to electoral political parties. Framing a message, building support and acting strategically to exploit opportunities are important parts of political action in general, as well as specifically for social movements.

Many forms of social movement activity aim to appeal to the state, but the state is not unambiguously our friend. There are Marxist and Anarchist critiques of state power, which suggest that states work not for the common good but to protect elite interests (Lenin 1965; Kropotkin 1943). Equally, independently of these, even if the state is assumed to be neutral, unintended consequences can arise from any policy proposed. So social movement activity that instructs the state to introduce policies appropriate to an emergency is likely to lead to a number of negative consequences. This is discussed in more detail in Chapter Nine.

There is, I would suggest, both a totality and particular points of potential engagement within it. A capitalist mode of production, in the totality of its processes and internal relationships, tends to degrade the environment. Therefore we need strategies to move us beyond capitalism. But a mode of production, or if you like, a society, while a totality, is also a network of different inter-locking forces, processes and structures. These can be examined, and interventions made. For example, in the UK we have a history both of imperialism and finance. British banks are an important source of funds for global fossil fuel projects; intervening around this is a specific but achievable tactic. Likewise, militant particularism, passionate local campaigns to protect the environments that communities love, can contribute to global climate change action. Opposition to fracking and open cast mining are relevant here.

XR, at the time of writing, has done a good job in raising awareness of climate change. Such awareness, on the one hand, must be translated into specific achievable victories: the closure of an open cast coal mine, the prevention of a new motorway link or the fight against fossil lending from a British bank. More radically it needs to shape strategies that moves us beyond capitalism. Such strategic anti-capitalism has been a part of the most militant climate change movements. A creative tension between immediate gains in halting climate change emissions and more long-term fundamental transformation is necessary.

The next chapter will discuss right-wing counter movements and the notion of social feedback. Climate change produces insecurity and right-wing activists exploit this to gain more influence, both by electing right-wing politicians and by shaping a culture that rejects concern around climate change. Such social feedback is dangerous and, like natural feedback, it has the potential to accelerate catastrophe.

Chapter Seven

The Climate Accelerationists

> In the field of social ecology, men like Donald Trump are permitted to proliferate freely, like another species of algae, taking over entire districts of New York and Atlantic City; he 'redevelops' by raising rents, thereby driving out tens of thousands of poor families, most of whom are condemned to homelessness, becoming the equivalent of the dead fish of environmental ecology. (Guattari 2014: 28)

The 2019 Australian election was labelled by commentators as 'the climate change election' (Morton 2019). Raging forest fires and droughts drove concern. The Australian Labor Party, which had been ahead in the polls, promised to take serious action to reduce carbon emissions. The result, shocking many pollsters, commentators, and above all those concerned with protecting our planet, was the re-election of the governing Liberal-National coalition.

As in an increasing number of countries, this right-wing coalition has no interest in tackling climate change. Indeed 'Vote for us and temperatures will rise even further', might have been their slogan, or 'Trust us, we will increase fossil fuel production and ruin your lives'. While green, left and otherwise radical climate change campaigners have, in previous decades, focused on the deficiencies of centre ground attempts to deal with the crisis, such as carbon pricing, an aggressive political force promising to rip up even modest policies to protect the environment has grown in strength and power. A key task, if we are to put out the fires, is to understand the growth of right-wing climate change denial, so as to challenge it more effectively. We might even term many of those promoting denial as 'climate accelerationists'.

Of course, I am being a little polemical to say that they are conscious accelerationists. Yet I have an image in my head of Donald Trump closing a deal with extra-terrestrial land speculators, promising them that he will dig so much coal out of the American ground that temperatures will become fiery enough on our planet to make it pleasant for pensioners from Venus to purchase holiday homes here. The enthusiasm for promoting fossil fuels and making a bonfire of any environmental legislation can appear rather bizarre. The Australian experience is instructive here.

Scott Morrison, the Liberal Party MP, cabinet member and Treasurer of Australia from 2015 to 2018, brought a large lump of coal to the House of Representatives. He taunted the opposition Labor Party leader Bill Shorten, insisting that an obsession with renewable energy was responsible for recent power cuts, 'Don't be afraid, don't be scared, it won't hurt you. It's coal.' The pro-coal coalition Prime Minster Tony Abbott was forced to resign, but his place was taken by Morrison. Labor, wounded by their defeat in the Federal election, have since weakened their climate commitments and it looks likely that the huge Adani coal mine will be opened (Griffith 2019).

While Abbott lost his Warringah constituency in the Federal election and the combined Green-Labor vote was a little higher than the pro-fossil fuel coalition, rising temperatures and damage to the Barrier Reef have not translated into a popular majority commitment to tackle climate change in Australia. Pro-fossil fuel billionaires and corporations have bankrolled campaigns for coal, but cash is perhaps not the only reason for the success of a pro-fossil fuel political right.

In Europe, several states are also led by governing parties with an affection for coal. In June 2019, Hungary, Poland, Estonia and the Czech Republic removed an EU commitment to reduce emissions to net zero by 2050. Individual EU states can often veto EU policies, so these four states blocked the proposals that had been endorsed by the other members (Keating 2019). In Poland, coal has made a historic contribution to the economy, while miners symbolise working-class heroism. Perhaps ironically, the 2018 UN climate conference COP24 was held in Katowice, a city known as Poland's coal capital. The Katowice event proved to be a festival of climate accelerationism, as coal lobbyists promoted fossil fuel extraction:

> Even the slogan of COP24, 'Katowice is changing the climate!', which is displayed on buses, has a sense of black humour to it. [...]
> The country's stand is covered with coal and displays products derived from coal, such as soaps and jewellery. On 4 December, at the opening press conference, Polish President Andrzej Duda highlighted the 200 years' worth of coal reserves the country possesses. (Robert 2018)

Countries such as Poland and Hungary are led by individuals who are generally classified as right-wing populists; their hostility to renewable energy is combined with fierce opposition to refugees and other ultra-conservative politics. The right-wing German Alliance for Germany (AFD), while at present a minority party distant from government, is another example of a European political force hostile to climate change action. The AFD's co-leader Alexander Gauland asserts that the 'Greta cult is reminiscent of collective hysteria in the Middle Ages', while predicting that renewable energy would turn Europe into a 'deindustrialised settlement region covered in wind farms' (*France24*: 2019).

Perhaps the most extreme manifestation of right-wing opposition to climate change action and support for renewables comes from the US Republican Party. President Donald Trump's accelerationism is largely the norm for most elected officials in his party. Fervent denial is a mark of political orthodoxy on the

American right. Since as far as back as Ronald Reagan's victory as President in 1980, Republicans have often campaigned against reducing emissions. Trump has pulled the US out of the Paris Agreement on climate change, enthusiastically worked to support the coal industry, appointed a climate change denier to head the Environment Protection Agency and even removed the term 'climate change' from government websites. In 2018 he commented that huge forest fires in California were a result of local mismanagement and had nothing to do with climate change:

> There is no reason for these massive, deadly and costly forest fires in California except that forest management is so poor [...] Billions of dollars are given each year, with so many lives lost, all because of gross mismanagement of the forests. Remedy now, or no more Fed payments! (Sommerlad 2018)

Policies to make it more difficult and expensive to install renewable energy have been another area of accelerationism. Tariffs have been imposed on imports of solar panels by Trump, although of course his protectionist sentiments don't stop with renewables. The Republican Party seems committed to making life difficult for cleaner energy, in Nevada in 2015, a law was passed allowing electricity companies to charge extra to consumers who installed solar panels. In a state with a lot of potential for solar energy, the solar industry stagnated and three major companies, SolarCity, Sunrun and Vivint left the state. Thousands of jobs were lost as a result in Nevada (Ryan 2017).

Another climate accelerationist is Jair Bolsonaro, elected in 2018 as Brazilian President. As in Australia and the USA, mounting evidence of extreme temperatures that translate into real human suffering are combined with an enthusiasm amongst major segments of the electorate with a politics that apparently aims to make things worse. Bolsonaro is openly homophobic and misogynist. He praises Brazil's former military dictatorship, supports death squads and advocates taking land from Indigenous communities. At my most pessimistic, I have often wondered whether his victory was a defeat for the whole of the human race. Certainly, his potential to accelerate climate change is strong given the importance of the Brazilian part of the Amazon. Supported by Donald Trump, Bolsonaro promised to take Brazil out of the Paris climate change agreement; since his election he has proved unable to do so. However, killings in the Amazon and deforestation have quickly accelerated; Bolsonaro, in true Trump style, is weakening environmental legislation. In August 2019, the world reacted in horror as the Amazon burned after Bolsonaro had encouraged loggers and cattle ranchers to destroy it. Like Trump, he is a master of the absurd, typically accusing the environmentally concerned activist and actor Leonardo DiCaprio of helping to set fire to Brazil's rainforests (Neale 2019).

There is a danger that climate change denial will become the consensus position amongst not just far right but traditional conservative, Christian Democrat and other centre right political parties. In Britain, the rise of the UK Independence Party and the emergence more recently of the Brexit Party shows new right-wing parties, hostile to climate change action and with strong links to Trump and other far right leaders, gaining votes. In 2019 the Green Party did well in the European Elections, moving from three to seven members of the European Parliament (MEPs), but the Brexit Party came first with 29 seats. Many of their MEPs are climate sceptics; their success seems to be pushing the presently governing Conservative Party to the right. Of course in 2020 Britain finally left the EU and the MEPs lost their jobs. So far, the new Conservative Prime Minister Boris Johnson seems reluctant to join the Bolsonaros and Trumps but this could change. Matt Hope, editor of *Desmog UK*, which campaigns on climate issues, has warned that Johnson could rip up EU climate commitments, noting, 'contrarian climate arguments fit well with mainstream conservative parties' appetite for regulatory rollbacks,' and suggesting that Johnson might be tempted turn post-Brexit Britain into a loosely regulated, low-tax haven. As such, environmental standards would be cut. Britain too might make a trade deal with the US that could, given the strength of their climate accelerationists, further erode standards (Gardiner 2019).

The accelerationists are not in charge of climate policy in most parts of our planet, so we should not despair. In the USA, while the Republicans are opposing climate change policies on a national basis, many states and cities are introducing serious policies to reduce emissions. The rise of the Greens and the efforts of XR are reducing the advance of climate accelerationists in Europe. In Western Europe, climate concern is stronger than in the USA but populists and sceptics exist, and have the potential to shape how we respond to the threat of rising temperatures. If we are to successfully challenge them, we need to understand their origins, inspiration and arguments. Who are the climate accelerationists and what is the nature of the extinction lobby we face?

The extinction lobby?

The accelerationists have been stimulated and sustained by fossil fuel corporations since the 1970s. Firms involved in the extraction of coal, oil and gas have spent billions trying to deny the threat of climate crisis. For decades, companies including oil producers have known that their day to day business was fuelling climate change but, of course, rather than changing their business model, they have attempted to silence critics and block moves that might reduce fossil fuel use. Oil companies have variously funded scientific research to suggest that climate change is not happening, that it doesn't matter, or is caused by other factors than burning fossil fuels. Think tanks and other institutions advocating denial have spent millions of dollars and it is claimed that much of this money came from ExxonMobil Corporation. Between 1998 and 2005 Exxon provided,

> $16 million to more than forty organisations that deny global warming. The American Enterprise Institute (AEI) received $1,625,000; Lee R. Raymond, ExxonMobil's chairman

and CEO, served as vice chair of AEI's Board of Trustees. In February 2007. AEI wrote to scientists and economists in Britain and the United States offering each $10,000 for articles criticizing the IPCC's Fourth Assessment, which was about to appear. (Powell 2011: 110)

Tobacco companies in the 1950s and 1960s, finding that their product killed, spent large amounts of money using the very real scientific complexity apparent around the causation of cancer as a means of ignoring the problem. As a direct result perhaps hundreds of millions of early deaths occurred. Many of the same lobbyists and scientists hired by the tobacco firms have gone on to work for big oil and big coal, yet again sowing doubt and slowing positive change (Oreskes and Conway 2012).

Coal and oil companies are also working hard to help elect politicians who oppose renewable energy and support fossil fuel extraction. Exxon, for example, spends heavily on social media to campaign in US elections to defeat candidates who take climate change seriously. The Republicans might as well be owned by companies such as Exxon. Indeed, Donald Trump appointed Rex Tillerson, chair and chief executive of Exxon between 2006 and 2016, as United States Secretary of State, one of the most powerful positions in government. The Democrats are also heavily funded by fossil fuel corporations. While some Democrats are serious about climate change, the issue has proved challenging for them. A national debate in the party about climate change in 2019 was pulled amongst some controversy. It has also been revealed that Democratic Party events are often supported financially by oil companies and lobbyists, which might have influenced this decision:

> American Petroleum Institute (API), the top trade association for big oil and gas companies, stepped up with $700,000 for Democrats' last convention in Philadelphia. That fell just a few money bags shy of the $900,000 API sent to the GOP. (Symons 2019a).

However in recent years, with concern over climate change rising, the fossil fuel corporations have found it more difficult to directly fund denial. As I write, a threefold lobbying policy seems to have emerged (InfluenceMap 2019a). First, straightforward and unambiguous accelerationist denial is outsourced to third parties. In turn, millions are spent trying to delay or dilute climate policy. Finally, the companies, while expanding investment in fossil fuel extraction, greenwash their record by exaggerating their embarrassingly small spending on renewables.

Thus Exxon no longer challenges the notion that CO_2 emissions are causing rising temperatures (Breslow 2012). Yet they continue to indirectly fund carbon contrarianism via their membership of the American Petroleum Institute (American Petroleum Institute 2020). Even the API has softened its rhetoric on climate change but opposes most regulation; for example, they have supported Donald Trump's attempts to scrap rules to prevent leaks of methane (Piper 2019).

Delaying and diluting carbon reduction policies remains a major objective for big oil. Fossil fuel lobbyists argue that renewables are unreliable and reducing extraction will lead to unemployment. According to InfluenceMap (2019b) the largest five privately owned oil extraction companies, Exxon, British Petroleum, Royal Dutch Shell, Chevron, and Total, spent £1bn on lobbying in the three years after the 2015 Paris Climate Conference. All five spend heavily on marketing aimed at promoting the narrative that fossil fuel production must be expanded. In turn, perhaps confusingly, these companies argue that they are strongly promoting renewable energy. The defensive marketing rhetoric does not measure up to the reality. In recent years 97 per cent of the five's capital investment has gone into developing oil and other fossil fuels, with just 3 per cent spent on renewables. Exxon's stated goal of producing 10,000 barrels of biofuel based oil by 2025 is equivalent to just 0.2 per cent of spending on refinery capacity. This figure is essentially 'a rounding error' (InfluenceMap 2019b: 13). And of course biofuel has, to date, often proved to be a problematic form of energy (Wall 2009). Exxon's photographs of a second generation of algae based oil is used to sell their new image as a climate conscious company. Previous decades of funding direct denial and continuing donations to industry lobbyists mean that oil companies have already massively slowed action on climate change. The coal lobby, at present, is far bolder in its denial but pushes the 'clean coal' trope as a way of suggesting that they do care about the environment. Donations to shape electoral campaigns are likely to slow any action on climate, especially in a US context. During the 2018 US midterm elections, oil companies spent $2m on Facebook and Instagram political adverts in just four weeks (InfluenceMap 2019: 2).

How scientific is science?

Climate denial, as previously noted, is possible partly because the science of climate change is complex. It invites doubt because numerous interacting factors can, at least potentially, lead to climate change. More fundamentally, the practice of science is far less certain than those of us who are not scientists generally assume. 'Science' may not be the new religion, but like 'God' or 'Nature', it is often seen as an argument closer, a way of establishing without doubt what is correct, a means of closing an argument and ending further debate. Today, science is often conceived as providing a definitive answer to many questions. When it comes to climate change, it might be assumed that science with a capital 'S' provides an answer. Yet a wide range of studies putting forward an array of perspectives suggest that science is a complex field of knowledge construction and is subject to new perspectives. The history of science is not as varied perhaps as the history of art, but it does have a history and is subject to change. Both the sociologists of science and scientists themselves often put doubt at the centre of the research process. The rest of us often assume, perhaps, that certainty is the universal product of successful science. Scientific knowledge can be conceived, wrongly in my view, as a black box. We may not know all the steps in the process, but we can be certain that the box exists and provides an answer, even if we have little or no idea what goes on inside the box. Therefore, any more sceptical approach to science can have a devastating effect on our trust in it.

The ever provocative French thinker Bruno Latour challenges such a 'black box', or what he terms an 'Immaculate Conception', view of science (Paulson 2018). Latour, via his actor-network theory, asserts that the production of scientific knowledge is a product of alliances, struggles and accidents. To the dismay of his critics, he shines a light on the often dingy and contested process of assembling scientific knowledge. Latour, while controversial, is just one of many philosophers, sociologists and other theorists, who note that science is a product of a contested cultural process. Perhaps best known is the American physicist Thomas Kuhn, who wrote *The Structure of Scientific Revolutions*, arguing that an accepted scientific view, a paradigm, often collapses under the weight of new evidence and is replaced by a new perspective (Kuhn 1992). Typically, Isaac Newton's physics was replaced by that of Einstein's, while Ptolemy's view of the solar system was destroyed by the Copernican revolution.

Another philosopher of scientific method, whom I have found instructive, is Gaston Bachelard. A former postal worker turned philosopher, he argued that science proceeds, as suggested by Kuhn, through revolutions. Knowledge does not slowly accumulate, and science effortlessly evolve as a result. New facts and new experimental results might be ignored until they build up to such a degree that a previous theory is swept away, as a dam might be swept away when the force of the water accumulating from storms becomes so strong as to destroy the previously secure structure. Bachelard's particular point was that science often involves research into phenomena which are not immediately visible. None of us can see atoms or molecules directly, so some kind of pictorial image may be used to represent them. Such images can be extremely misleading and often are discarded, but may be necessary to the development of new scientific theories. The term 'epistemology' means the study of knowledge, Bachelard argued that breaks and sudden changes in epistemology often occur. Aware of the 'false union of the concept and the image', he argued that a sudden change in knowledge, the epistemological break, occurred when we discard a particular image used to imagine how a concept worked and adopted a new image for our investigation (Smith 2016: 120). Bachelard's rigorous approach was inspired partly by Cavaillès (Ferriéres 2003: 137).

Diverse perspectives investigating scientific method also include the anarchist 'anything goes' approach of Paul Feyerabend who wrote *Against Method* (1975). All these approaches challenge positivism, and we can add more, from Karl Popper's arguments concerning falsifiability to the 'strong programme', which examines the social nature of science (Nola and Sankey 2007). While most of us don't use the term, I suspect positivism expresses how we non-scientists generally think science works to provide facts that are strongly founded. Positivism is a philosophy of science that argues that absolute knowledge can be established through observation backed up by logic. Science produces, it is thought, certainty. Critics of positivism argue that this is too simplistic; scientific knowledge changes, it often does so by way of dramatic transformation, and how we investigate reality is shaped to some degree by social context.

While they don't engage in detailed discussion of the philosophy or sociology of science, Oreskes and Conway (2012) in *Merchants of Doubt*, an examination of climate change denial, note that we often tend to think science is about 'facts', but they agree instead that doubt drives research. Deniers exploit such doubt to attack any kind of conclusion derived from research that suggests CO_2 leads to climate change. Normal science is a surprisingly messy pursuit.

While science isn't by any means simple, even the strong programme sociologists will, if pressed, agree that it isn't entirely a random artefact of human society, we can make some progress in knowledge. For example, it is unlikely that new discoveries will take us back to a view that the Earth is the centre of the solar system as Ptolemy believed or that Newton had a more sophisticated understanding of physics than Einstein. Climate deniers tend to weaponize doubt. Therefore, anything but what-you-see-is-what-you-get certainty can be undermined. And science is far less certain than we might generally think. Perhaps science has always been to some extent political; certainly religious authorities have seen new science as a threat to their authority. Galileo was persecuted by the Catholic Church. The 1925 Scopes Monkey trial is an example of the threat to theology and, in turn, political authority felt by some from the teaching of evolution (Sprague de Camp 1968). The various sociological approaches from Latour and others don't however tend to suggest that science is usually a political battle. Climate change denial can be seen as a continuing attempt to use varied tools to challenge arguments that climate change is caused by human action. The realisation that science is less certain than we might think, can, of course, be used by the climate acceleration lobby, to promote doubt and inaction. Thus sophisticated attempts to express the complexity of scientific method might be exploited by deniers.

A good example of this is cited in *Merchants of Doubt* around the question of fingerprints. One area of extensive discussion, which I have described briefly in Chapter Two, is the fact that different factors can cause climate change. Volcanoes, sunspots, atmospheric dust and water vapour may influence climate, but CO_2 emissions and various changes in the Earth's orbit (Milankovitch cycles) are generally seen as the most significant factors that can cause major temperature changes. One approach to deciding which matters the most is to look for 'fingerprints'. Thus, it is argued that solar activity will heat the entire atmosphere, but CO_2 emissions will have different effects at different points in the atmosphere. Thus, a rise in temperature because of solar activity will have a different 'fingerprint' to a rise caused by increasing emissions.

Benjamin Santer, who worked as an atmospheric scientist at

the U.S. Department of Energy's Lawrence Livermore National Laboratory during the 1990s, researched satellite data on such fingerprints. Solar driven climate rises would, it was assumed, show up most strongly in the upper atmosphere (stratosphere). Climate change driven by CO_2 would show up more strongly in the lower atmosphere. Santer established that the lower atmosphere (troposphere) was indeed warming substantially. This was a conclusion that was uncomfortable to the climate change deniers:

His work found that while the troposphere was warming, the stratosphere was cooling, which was a strong form of fingerprint evidence for CO_2 driving climate change. In fact, with climate change, the boundary between the troposphere and the stratosphere was rising, indicating that the atmosphere as a whole was radically changing. He contributed to the IPCC 1995 report which argued that human action was driving rising temperatures. A group of physicists associated with a Washington, D.C. based think tank argued that Santer had distorted the report, ignoring dissenting voices. They accused him of 'scientific cleansing' (Oreskes and Conway 2012: 3). They wrote to members of Congress, produced documents with titles such as 'Greenhouse Debate Continued' and 'Doctoring the Documents' and sent off numerous letters to both newspapers and scientific journals alleging him of malpractice. They put pressure on his employers to fire him. If we fast forward to 2020, sceptics and contrarians now rarely dispute the satellite data used to show the fingerprint, and have moved on to other issues. In the 1990s this data was strongly contested and the attacks on Santer and others became personal, driven via media rather than scientific channels:

> In 1996, at the time of publication of the IPCC Second Assessment, I was a messenger bearing news that some very powerful people did not want to hear. So they went after the messenger. They were very good at it. I'm sure there was no personal animus involved. I just happened to get in the way and had to be discredited. (Santer in Thacker 2006: 5837)

The physicist Frederick Seitz wrote an editorial piece for the *Wall Street Journal* entitled 'A major deception on "Global Warming"', stating that significant changes in Chapter Eight of the IPCC report had been made and that Santer as the lead author must take responsibility,

> IPCC reports are often called the 'consensus' view. If they lead to carbon taxes and restraints on economic growth, they will have a major and almost certainly destructive impact on the economies of the world. Whatever the intent was of those who made these significant changes, their effect is to deceive policy makers and the public into believing that the scientific evidence shows human activities are causing global warming. If the IPCC is incapable of following its most basic procedures, it would be best to abandon the entire IPCC process, (Seitz 1996)

Responding, Ben Santer noted that Seitz was neither a climate scientist nor had any direct involvement with the IPCC process. Seitz' criticism, according to Santer, was similar to that of the Global Climate Coalition, an oil industry body who 'conjure visions of sinister conspiracies that are entirely unfounded.' He further noted:

> There has been no dishonesty, no corruption of the peer-review process and no bias -- political, environmental or otherwise. Mr. Seitz claims that the scientific content of Chapter 8 was altered by the changes made to it after the Madrid IPCC meeting. This is incorrect. The present version of Chapter 8, in its Executive Summary, draws precisely the same 'bottom-line' conclusion as the original Oct. 9 version of the chapter: 'Taken together, these results point towards a human influence on climate.' (Santer 1996)

Santer was subject to vicious personal abuse, a dead rat was dumped on his doorstep, and attacks labelling him biased and unprofessional have continued for decades (Santer 2019).

From cigarettes to climate change

In 1953, scientists had demonstrated that cigarette tar painted on to mice caused cancer. This was something of a blow to the major tobacco firms (as well as the mice used for research) and they swiftly organised to challenge the science. According to *Merchants of Doubt,* the four CEOs of the largest tobacco companies met and agreed to convince the public that there was no scientific basis to claims that their product was carcinogenic (Oreskes and Conway 2012). Cancer, like climate change, is complex; many smokers never get cancer and many of those with lung cancer have never smoked. There is still today some debate about why smoking has differential effects on different individuals (Ryan 2018). Correlations don't always neatly transmit into causation. Rising temperatures and rising CO_2, or increased smoking and increased cancer, demanded investigation. In both cases, industry has often attacked research that might show causation rather than investigating the possibility seriously.

The Tobacco Industry Research Committee was established, by major cigarette companies, to promote research challenging the argument that the product causes cancer (Oreskes and Conway 2012: 15). Scientists were found, research grants allocated, and doctors lobbied. Contact was made with newspapers and magazines like *Reader's Digest*. Opinion polls were used to gauge public attitudes and provide data with which to reshape them. In January 1954 'A Frank Statement to Cigarette Smokers', a full page advertisement, was placed in more than four hundred US newspapers, by the Tobacco Industry Research Committee, reaching approximately 43 million Americans. Made to look more like an editorial than an advert, and personally signed by major tobacco industry figures, it made the following central claims:

> That medical research of recent years indicates many possible causes of lung cancer.
> That there is no agreement among the authorities

regarding what the cause is.

That there is no proof that cigarette smoking is one of the causes.

That statistics purporting to link cigarette smoking with the disease could apply with equal force to any one of many other aspects of modern life. Indeed the validity of the statistics themselves is questioned by numerous scientists. (Sourcewatch 2009)

There has been considerable overlap between the key individuals promoting doubt about tobacco and cancer, and those challenging the link between fossil fuels and climate change. Seitz, as well as later opposing the science connecting carbon emissions to climate change, was a permanent scientific advisor to R.J. Reynolds Tobacco Company between 1979 and 1988. He was no fringe figure, but one of the USA's top scientists. A physicist, he wrote *The Modern Theory of Solids* which was published in 1940. He was director of atomic training at Oak Ridge National Laboratory from 1946 to 1947. His career seems to be one of impressive advance, by 1962 he served as President of the United States National Academy of Sciences a post he held until 1969. Between 1968 and 1978 he was President of Rockefeller University. Working for R. J. Reynolds, between 1978 and 1987, he organised $43.4 million in research grants (Oreskes and Conway 2012: 29). While not all of this directly involved disputing the science around cancer and tobacco, much of it did. Links with research scientists, both direct and indirect, were useful to the tobacco industry who needed sympathetic witnesses as litigation cases mounted. Stanton Glantz, a professor of medicine at the University of California, observed that the research Frederick Seitz oversaw for Reynolds was,

> perfectly fine research, but off the point [...] Looking at stress, at genetics, at lifestyle issues let Reynolds claim it was funding real research. But then it could cloud the issue by

saying, 'Well, what about this other possible causal factor?' It's like coming up with 57 other reasons for Hurricane Katrina rather than global warming. (Hertsgaard 2010)

Seitz established the George C. Marshall Institute in 1984. From promoting increased spending on nuclear defence to challenging the idea that the ozone layer was thinning because of CFCs (Chlorofluorocarbons), the Institute acted as an energetic contrarian research centre. Increasingly, Seitz moved from defending tobacco to challenging the science on climate change. Along with criticising Santer, as described earlier, Seitz promoted the Oregon petition, collecting signatures from individuals who challenged mainstream climate change science. Despite Seitz being a former President of the organisation, the National Academy of Sciences issued a disclaimer rejecting the petition. While thousands of individuals had signed, only a fraction had been actively involved in climate science (Oreskes and Conway 2012: 245).

Another key figure has been Fred Singer. As a research scientist he produced papers on physics and geophysics during the 1950s and 1960s. After 1970 he seems to have largely ceased taking part in research and instead branched into advocacy. He was a key critic of the notion that second-hand smoking caused cancer, publishing *The EPA and the Science of Environmental Tobacco Smoke* in defence of the industry (Oreskes and Conway 2012: 270). Described as 'the grandfather of global warming denial', he warned oil companies that they could be put out of business (Powell 2011: 54). He assembled a wide range of lobbyists including asbestos advocates to attack the US Environmental Protection Agency. One of the most scientifically qualified of the deniers has also been one of the bluntest, stating 'There is no proof at all that current warming is caused by the rise of greenhouse gases from human activities.' (Powell 2011: 58)

Thinking the unthinkable

Conservative think tanks, mainly US based, have worked hard to maintain doubt in climate change science (Dunlap and Jacques

2013). While corporations have funded scientific research to help nurture doubt, the growth of a consensus around mainstream climate science has seen fossil fuel producers increasingly move towards more subtle forms of lobbying. However, climate change denial has become an essential component of right-wing identity politics in a number of states, but especially the USA.

It has become an article of faith in the Republican Party that climate change caused by CO_2 emissions is a 'hoax'. Even when particular tropes and challenges are rebuffed by scientists, they often maintain a zombie existence on social media and amongst journalists. While the science, despite the complexities of climate change, has become more certain amongst academics, resistance has gained strength. We cannot attribute the success of climate accelerationism purely to corporate power and supposedly corrupt individuals. Perhaps it is a little more complex than a description of good climate activists, disinterested scientists and evil deniers.

Social feedback

Climate accelerationism represents a form of social feedback. Natural feedback can, as noted, be positive or negative. Positive leads to greater change, negative restores balance. Our bodies become warm, we sweat, this is negative feedback and keeps our system stable. Temperatures rise, more rock is revealed as ice and snow melt, the darker surface of the rock absorbs more sunlight, which means temperatures rise more, with such positive feedback from the albedo effect. We are in an era of positive social feedback: temperatures rise, more right-wing climate deniers are elected to office, they promote fossil fuel production, temperatures rise further. It is as if we were being propelled on an escalator to hell. The analogy breaks down as the escalator, rather than descending smoothly, looks in danger of speeding up. Trump, Bolsonaro, Scott Morrison with his lump of coal; hell is next on the list.

The most basic insights of Freud might be applied to such positive feedback. As the danger rises, for some of us, denial becomes stronger. It can seem absurd that as the science becomes clearer and the symptoms of crisis more obvious, deniers shout louder. Climate change, curiously, leads to greater rejection of the need for climate change action, at least for some members of society. Climate change, along with rising inequality, technological change and the 2008 financial crisis, means that life is becoming less certain and less predictable. This may have created a crisis of political representation, with established political parties in many countries collapsing or being transformed. Part of this process has included the often sudden rise of the right-wing populists. Bolsonaro was a minor political figure in a tiny party at the fringes of the Brazilian political system just a few months before his victory in 2018. The election of Trump as President was a joke in the long running cartoon show *The Simpsons* from as far back as 2000.

Perhaps without studying in depth the politics of individual countries such as Australia or Brazil, it is difficult to analyse such feedback. With climate and other forms of change, uncertainty rises, the change is based on complex factors that are often rather abstract. Political entrepreneurs on the new populist right exploit popular fears to enhance the visibility of 'the other'. A hate figure based on racism or other forms of discriminatory prejudice is framed. Pre-existing tropes are polished, reshaped and made more visible. Imaginary solutions to real problems are proposed, but the real may be invisible and the imaginary may be made clear and bright to catch the imagination of a significant groups of voters. Fake news may be part of the process, but it is used to nurture forms of fear that already exist; regulating social media might reduce amplification but would not cut the roots of this new denialist right. Trump focused on the figures of the Mexican, the migrant and the Muslim. In Europe, the crisis in Syria led to refugees seeking sanctuary. This increase in refugee populations across Europe was exploited by a host of far-right populist politicians from Britain through to Poland, leading to major political change.

Climate change will lead to more refugees. Movements of people will continue to be framed as a source of danger by populist right-wing politicians; they may well gain greater electoral support as a result. As they gain more power, they are likely to work to accelerate climate change, by attacking renewables, promoting coal and smashing environment regulation. This will generate more disasters, which will in turn generate more refugees. An increase in the number of refugees will be exploited by the right-wing politicians to enhance or maintain their power. This tendency is not universal, for example, Modi in India has, as a far-right leader, mobilised against the country's Muslims, but he has not been a significant advocate of climate change denial. Nonetheless in many countries the basic cycle of positive social feedback is clear. Climate change becomes a greater and greater problem, we know that it is affecting us now, but as it rises in significance, those dedicated to making it worse may gain more power. Challenging climate accelerationism, a noxious form of positive social feedback, is an urgent task.

It is important perhaps to name an enemy, but it is also important to remember that those one opposes generally argue that they are the children of good rather than evil. The story of Seitz and Singer is bemusing. Serious scientists with long careers, they became partisans in a corporate war against serious research. Continually they attacked parts of the research agenda; when the attack was repulsed in one area, they moved on to another. Disconcertingly, they used the media to aggressively criticise fellow scientists. The slowing of action on both tobacco and climate change are both causes of human suffering and in many cases death. There is a logic, however, to their actions. Both were committed to notions of the benefits of free markets and technological advance. While Seitz and Singer were paid for their work to nurture doubt, one gets the impression that this was not their primary motivation. They were enthusiasts rather than mercenaries. To a large extent both were cold warriors, the peak of their mature career occurring during the conflict between the Soviet Union and the USA that dominated international events from the 1950s to the 1980s. The science of smoking or climate change is a secondary issue, these are just manifestations of attempts to challenge technological advance and the free market. A right-wing, free market-based approach, fearful of big government, sincerely held, is part of the explanation for the persistence of contrarian views on climate change. As such, even as the scientific arguments are dismissed, the greater discussion is ignored. It is easy to demonise the other, but understanding how the others operate, their assumptions and methods, helps us more effectively challenge their works. Both climate campaigners and contrarians view themselves as on the side of what is good; the contrarians don't celebrate evil as an abstract and beautiful force, even though they may praise coal.

Climate apartheid

Climate change caused by rising emissions is often bound up with racism. I have noted already that the imperialist roots of a fossil fuel economy are well described by Andreas Malm (2016). The term 'climate apartheid' has been coined to describe a likely future, where richer and whiter populations continue to prosper even in the midst of apocalypse, while the rest of humanity finds life hotter and more difficult (Anon 2019c). The concept of environmental racism is important here. It can be argued that those who are whiter and better off can protect their environments but that pollution and other forms of damage are displaced to areas where minorities live. It has been argued, for example, that African Americans are more likely to see their communities blighted by smelters, garbage dumps and incinerators, because those with more influence can stop them being set up next to their homes (Taylor 2014). Coal and oil extraction are more likely to impact on poorer populations and they are less likely to gain benefits from such extraction.

Climate apartheid is tightly bound up with denial and social feedback. Those who can protect themselves better from rising sea levels or extreme weather care less about climate change. As populations are displaced by climate change, anti-migrant populists will do all they can to prevent them moving. While the connections are not universal, climate denial may correlate with racism. Again, the Trump experience is instructive, he has put huge energy both into attempting to accelerate climate change and building a landscape of repression, fighting hard to build a wall between the USA and Mexico while investing in a brutal regime of camps in which to imprison migrants. Yet it should not be forgotten that previous, apparently more liberal American Presidents, including Barack Obama, massively expanded this system of imprisonment too (Levitz 2016).

The Extinction Rebellion (XR) concept might fail to reveal that we are not all affected as one species uniformly by climate change. The division of race, class, gender and other power differentials mean that extinction, or at least accelerated misery, will be the fate of some but not all. It is likely that an ever-smaller habitable space will provide sanctuary for the richest and best connected, while the rest of us are left to drown, starve or burn.

Beyond populism

Before Morrison there was Tony Abbott. A former Australian Prime Minister, Abbott has proved to have had an enduring influence on the country's approach to climate change. His assertion that climate change can be dismissed as 'crap' and his framing of climate change policy as a carbon tax which would damage living standards, have endured. It has been suggested that his success in derailing action on climate change is a result of his impressive skills as a political communicator. Abbott's ability to communicate a narrative of climate denial rested on a number of key factors. These included strong message disciplines, setting the frame by defining potential climate change mitigation policies as a carbon tax and endlessly repeating the term. He baited the left by making provocative statements; this gained attention and made sure his message was endless repeated. This is also key in explaining Trump's success as a communicator: extremism gains attention, rebuttals amplify the original message rather than refuting it. Finally, Abbott worked communication magic for his contrarian message by focusing, on concrete case studies rather than abstract statistics and targets. Abbott and co. spoke about costs that were immediate and local — quarterly power bills, jobs in Australian towns and cities, the price of a lamb roast. (Foyster 2019)

The success of right-wing climate denying politicians, including Abbott, cannot be dismissed but must be challenged. Effective communication is part of such a challenge. Part of the war on climate, as has been discussed, is the ability of opponents of climate action to use huge amounts of cash to shape public perception so as to slow, and sometimes reverse, action on climate change. More detailed consideration of their strategies is necessary if we are to combat them. Images, narratives and such like have a material effect, shaping what we think and influencing what we do. Thus Chapter Eight examines climate change communication.

Chapter Eight

The Unconscious, the Imaginary and the Real

Gore's activism culminated in 2006, with the release of his documentary *An Inconvenient Truth*, but a month before its release, Comedy Central aired an episode of *South Park* ridiculing Gore as a hysterical buffoon obsessed with warning the nation about a mythological creature called ManBearPig, 'It is half man, half bear, and half pig.' […] We all had a good laugh at Gore's foolishness. After all, no one likes a complainer, a critic, a moralist, a killjoy, a naysayer, a bearer of bad news, and in fact a lot of us probably remember ManBearPig better than we remember *An Inconvenient Truth*. (Wynn 2018)

Perhaps it is not so surprising that we often ignore climate change. It can appear a distant problem, vast yet hardly visible. The climate accelerationists have made the imaginary threat of the 'other' in the form of migrants and other minority groups vivid and threatening. In contrast scientists may have rendered potential climate catastrophe sepia and far away. Communicating the threat of climate change has proved difficult on the whole, and while temperatures rise, public concern is often static. The opponents of any kind of serious action have made the supposed negative effects of reducing emissions more obvious, and the proponents have often struggled to have an impact.

Logical argument is, perhaps, rarely effective in changing minds, and the assumption that we make rational choices based on our own free will is open to question (Ravven 2013). Scientists have tended to work with an information deficit model, believing that climate change denial occurs because we don't have enough information (Norgaard 2011: 64). My suspicion is that in a world of messaging, we are captured by headlines, tropes, narratives and, above all, by emotionally charged images. Gaining a more sophisticated understanding of how we are shaped by such forces might just make us more able to decide what we believe. Professor Andrew J. Hoffman of the University of Michigan maintains that, fundamentally, our understanding of climate change is shaped not by science, but by culture (Hoffman 2015). In his book *Don't Even Think About It*, the environmental campaigner George Marshall goes further, arguing that our brains are hardwired to ignore climate change (Marshall 2014). Even without the malign efforts of the deniers, he believes, our basic mental make-up helps promote ignorance of climate change.

Sustaining the climate change unconsciousness

The psychoanalyst Donna M. Orange has suggested that we often banish the climate crisis from our conscious mind. We get on with our everyday lives, working, eating, going on holiday, perhaps flying, and forget the unfolding catastrophe. She believes that psychoanalysis can act as a tool to tackle this but those working with the unconscious can be just as unconscious as the rest of us. Orange notes that while psychoanalysts often perceive themselves to be intellectual leaders with a special insight into human motivation, they are often poor at examining their own motives. While the profession should perhaps have a prophetic function or, at least, be able to warn of dangers such as climate change, even Sigmund Freud seemed to have misjudged danger. Freud, apparently misled by his passion for his own working life in Austria, along with a love of German culture, was in denial about the threat he and his Jewish family faced from the Nazis in Vienna during the late 1930s,

He and his daughter Anna escaped to England [...] at the last moment, but several members of his family perished in the massacre. In another strange example, a few year later in the London Blitzkrieg, during one of the British Psychoanalytical Society's furious disputes about the origins of hatred and aggression, Donald Winnicott noted their actual effects: 'I would like to point out there is an air raid going on.' (Orange 2017: xii)

If disaster cannot be acknowledged, denial becomes ever deeper. Cognitive dissonance occurs so that even if we are conscious of climate change, we bracket it out, so that it won't influence what we do. Of course, despite criticism, this is one of the reasons why XR has been significant. Groups of people sat on the roof of our train carriage, threats of drones at the airport, and general inconvenience all gain our attention. One thinks of the prophets of the Old Testament; nobody much liked them, they were a raving inconvenience, but they warned the community of crisis. Freud and the British Psychoanalytical Society had a prophetic function, warning of the dangers of human aggression driven by destructive internal drives, but they overlooked the aggression that directly threatened them from Hitler's bombs.

Don't even think about it

The writer and environmental campaigner George Marshall has investigated the science of why we don't listen to climate scientists. Scientists present facts, but facts on their own don't grab our attention, overcome biases, or soothe doubts. He found that while evidence of climate change caused by carbon was rising, this did not necessarily correlate with increased public concern. Knowledge doesn't automatically translate into belief, let alone action.

In *Don't Even Think About It*, which is an engaging read, Marshall combines an examination of academic sources with interviews with various deniers and sceptics. He takes those he disagrees with seriously rather than demonising them. *Don't*

Even Think About It is both accessible to readers and based on strong analysis. Marshall's main insight, I feel, is that what we accept as true is what makes sense to us. Such sense is built from compelling and credible stories,

> I find that everyone, experts and non-experts alike, converts climate change into stories that embody their own values, assumptions, and prejudices. I describe how these stories can come to take a life of their own, following their own rules, evolving and gaining authority as they pass between people. (Marshall 2014: 3)

Above all, he notes, group identities shape what we take to be the truth. While Marshall is aware of the wall of money pushed to slow policy change on climate, he is also aware that it is insufficient to point to the malign work of corporations and conservative think tanks. Climate deniers are able to exploit features of how we commonly understand reality. He suggests too that the efforts of both advocates of action and their conservative detractors only capture the attention of a minority. Much of the resistance to taking climate change seriously and acting upon such an understanding is based on narratives and motivations which are too often ignored.

Marshall interviewed Daniel Kahneman, author of *Thinking Fast and Thinking Slow,* who won a 2002 Nobel Prize for his work on the psychology of decision making. Kahneman's work suggests that we humans have to process large amounts of information in our daily lives, so we use pre-established norms to make decisions. Such fast thinking is a short cut, which works to a certain extent but lacks sophistication and can be misleading. According to Kahneman (2012) we use mental cues, intuitive and inbuilt, based on previous experience, to make most decisions. What we think about something is, therefore, usually based on cognitive biases that enable the quick judgements necessary for fast thinking. This process may not be adequate for dealing with complex and entirely new situations. His research indicated broadly that people are more averse to potential losses than probably gains, more aware of short-term than long-term factors, and value certainty rather than uncertainty. Kahneman, when interviewed by Marshall, argued that climate change would tend to be ignored because it lacked salience. The most salient issues, Kahneman suggests, tend to be concrete, immediate and indisputable. Climate change fails in respect of each of these three criteria that are understood to make issues significant to human beings: it is perceived as a future problem, may not have immediately visible effects and is subject to much uncertainty. Asked about the likelihood of human beings adequately dealing with climate change, Kahneman was blunt: 'I'm extremely skeptical that we can cope with climate change.' (Marshall 2014: 57)

The painful paradox of ecology is that if a problem creates obvious effects it is often too late to do anything; we need to act before a problem is immediately visible. Climate change today is shaped by carbon emitted forty years ago. What we do now influences climate in future decades (Marshall 2010). Thus Marshall notes that climate change is a problem of the past, present and future. We have known about it for a long time (the past), its effects are becoming apparent now (the present) but will likely accelerate over time (the future). To gain our attention, campaigners have used the notion of time limits to try to promote urgency and to spur us into action. Marshall recalled that when he worked as an intern for the *Ecologist* in 1990 the magazine published a book entitled *5000 Days to Save the Planet*. Those 5000 days have come and gone. They ran out in the same year, he notes, that a British think tank, the Institute for Public Policy Research released a report entitled 'Ten Years to Save the Planet' (Marshall 2014: 61). Climate change is already with us already, such tactics appear misleading and can be ridiculed.

Extreme weather events are framed within systems of personal meaning, so even immediate destruction is open, perhaps surprisingly, to interpretation. Almost every day, as I write, extreme weather events are reported. In the last couple of days there has been astonishing flooding in New Orleans and storms have killed sixty people in Greece. None of these events can absolutely be linked to climate change, but we know that extreme weather is increasing and that climate change drives extreme weather events such as storms. However, as weather events become more damaging: more typhoons, droughts, dramatic falls of snow, a ratchet effect may occur as we become more familiar with such extremes and therefore begin to view them as a new normal. One chapter of Marshall's book describes conversations with members of a Texan community who lost their homes because of a heatwave that created huge fires. None of them associated their personal disaster with climate change.

Knowledge is accepted or rejected partly, perhaps even mainly, because of emotional associations. Providing climate sceptics with more information can on rare occasions work to change minds, but much research shows that more information may even reinforce scepticism. At one time, US Republican and Democratic Party voters with a knowledge of the basic scientific case were likely to have similar assumptions about climate change. Marshall suggests that as climate change became more contested between the two major US parties, 'the issue became polluted by political and cultural meaning'. (Marshall 2014: 124) The more Republicans knew about climate change the less they believed. Sceptics had, according to Marshall, a slightly better understanding of the science than those of us who see climate change as a real problem caused by rising emissions. Thus it might be argued that denial 'is due to a surplus of culture rather than a deficit of information'. (Hamilton quoted in Marshall 2014: 124) Confirmation bias is a powerful force. Given new information, we usually use it to reinforce pre-existing views. When given a choice between statements from different sources, we tend to listen to the source that coincides with our pre-existing point of view.

Marshall is also fascinated by socially constructed silence. We don't talk about climate change. When he dropped the topic into conversation he was often met with embarrassment by friends and family. He noted that the sociologist Stanley Cohen distinguished between ignorance and disavowal. Ignorance is when we don't know, while disavowal is the active choice not to know. Climate change disavowal may be produced when we know that activities we have a stake in (because they give us pleasure or are economically important) generate climate change emissions. Silences build on silences, according to Marshall, in a complex feedback loop (Marshall 2014: 89). Voters in focus groups are silent, so politicians are more likely to be silent, and organisations view climate change as something that does not concern them, so they are silent too. The silence has been broken by the likes of XR and Greta Thunberg, but the dangers of disavowal should not be dismissed.

Climate change, if it is to gain attention and be acted upon, needs to be part of an engaging narrative. Marshall notes that the writer Michael Crichton, well known for the television series *ER* and the *Jurassic Park* films, authored a novel promoting denial. Published in 2004, *State of Fear* tells the story of eco-terrorists in the Environmental Liberation Front who set off disasters so as to create mass panic about climate change. The intention is to use the panic to create an authoritarian green dictatorship. With a strong narrative, a struggle of good versus evil, and dramatic events unfolding, it proved to be a bestseller. Fiction, if it is compelling, can have a powerful influence on our perception of the truth. Marshall notes that when he was writing, Crichton's novel was easily the single best-selling book on climate change. Even though he had no scientific qualification, Crichton was called to give evidence on climate change by the US government (Marshall 2014: 108). Climate change is framed by deniers on the right as a means by which those on the left seek to gain more control over individuals. The contrarian right like to frame not just carbon taxes but almost any policy approach aimed at reducing emission as a costly assault on human freedom. Crichton's novel amplified and popularised such assumptions. Despite being entirely fictional, *State of Fear* provided a strong cultural frame to understand climate change, in this case as a hoax exploited by green extremists.

The transference of a scientific debate to a cultural argument linked to contending ideological positions is more common than one might at first assume. The Australian academic Clive Hamilton noted that in the Weimar Republic of 1920s Germany, physics became the subject of intense political debate. A right-wing social movement campaigned against Einstein's theory of relativity. Einstein was a left-wing Jew and, with his love of Spinoza, an advocate of heterodox religious views. Opponents argued that Einstein's physics promoted moral relativism. The Nazis agitated for a specifically Aryan physics in opposition to his work. Einstein, in some confusion, wrote to a friend: 'The world is a strange madhouse. Currently, every coachman and every waiter is debating whether relativity theory is correct. Belief in this matter depends on political party affiliation' (quoted in Hamilton 2013: 20).

Belief can be a product of battles over a single word. The Russian literary theorist Michal Bakhtin argued that the particular meaning of a word could act as an arena of political conflict (Holquist 1990). Words reflect particular assumptions and trigger specific associations. There has been a confrontation, for example, over the phrases 'climate change' and 'global warming'(Marshall 2014: 125). Hoffman suggests that in framing effective climate communication, attention to apparently 'simple word choices' is essential. He notes that words have multiple meanings for multiple groups of people and that they can trigger particular reactions, either intended or not, when used. In discussing oil extraction from the bitumen sands in Alberta, Canada, he observes,

> the term 'tar sands' signals that one opposes the process, while 'oil sands' signals that one supports it. The word 'green' can become divisive, [...] the word 'sustainability' has become politicised [...] When scientists talk of 'uncertainty,' they are referring to a measure of statistical deviation around a mean.

But laypeople hear 'uncertainty' and think that scientists 'just don't know.'(Hoffman 2015: 63-64)

Another area of importance is the existence of an economy of attention (Doran 2017; Lanham 2006). We see and hear millions of potential messages. Advertisers compete to gain our attention, and with the existence of social media, opportunities have multiplied to do so. Beyond all the other reasons discussed, a barrier to climate change communication may be that we just don't see the messages. There has been a major debate in recent years about the existence of apparently fake news, a personal favourite is the story propagated on Facebook that the Pope had endorsed Donald Trump during the bitterly fought 2016 US Presidential election (Gordon and Todorova 2019: 171). This was a plain lie. Claims, whether straight up lies, half-truths, or some other fraction, combined with emotive images, gain our attention and can influence how we act. Fake news is not a new phenomenon. We tend to frame messages into pre-existing understandings, so even false information is likely to be meaningful and to be absorbed, understood and transmitted, if it fits with our existing understanding. As the podcaster Natalie Wynn notes, 'we live in a marketplace of attention, not a marketplace of ideas'. (*Economist* 2018)

Belief is based largely on trust and trust is dependent as much on the messenger as the message. If we hear or see information from a source that we trust, we are more likely to believe in it. Trust is, in turn, a product, at least partially, of identity: we categorize ourselves within particular groups. We are most likely to believe something if we learn it from others with whom we feel a shared identity. We tend to react negatively to ideas communicated from a group we distrust. Both climate change campaigners and climate change deniers whom Marshall met showed these features. It was not that one group was apparently more likely to trust in logic or reject conspiracy theory than the other. We are largely captured by groupthink but, of course, may believe that we are more sophisticated and that such groupthink only influences our supposedly ignorant opponents. The political scientist Jeffrey Friedman suggests that we may adhere to webs of belief, where by the opinions of political opponents may:

> seem irrelevant, implausible or even incomprehensible. This, indeed, would appear to be the source of a great deal of the mutual vitriol that is a given in politics: the fact that those on the other side seem to have such plainly wrong or unfathomable ideas, prompting the thought that they are irrational or evil. (Friedman 2019: 235).

Marshall argues that enemy narratives are almost universal but also that we need to challenge them to deal with climate change. This fits with the discussion of framing outlined in Chapter Six; effective campaigning works if we have something to campaign for or against. We mobilise to challenge a group or institution which we blame, government or corporations being favourite targets. Equally we cherish what we agree with. The closer a statement is to our opinion the more we value it. The French novelist Marcel Proust put this well:

> He was, indeed, in the habit of always comparing what he heard or read with an already familiar canon, and felt his admiration quicken if he could detect no difference. This state of mind is by no means to be ignored, for applied, to political conversations, to the reading of newspapers, it forms public opinion and thereby makes possible the greatest events in history. (Proust 2000: 469)

The right-wing political theorist Carl Schmitt argued that most politics came down to a 'friend-enemy' distinction (Newman 2018: 4). If we replace enemies with friends by winning an election or making revolution, we might see such victories as signalling that the game is over and our goal achieved. The friend-enemy opposition is bound up tightly with collective and individual identity. This is a likely reason why it is so difficult to persuade others, why facts may not actually matter very much in debates and why political debates can become so aggressive. Political opinions become markers of identity. If we can disconnect friend-enemy distinctions, difficult as this may be, Marshall argues, we can make it easier to engage in discussion with others and perhaps change minds.

Belief is linked to particular practices that appear necessary to maintain identity. A collective culture that includes practices of driving large vehicles, aggressive meat eating, etc, may make a belief in climate change action difficult to sustain. If believing in climate change suggests the rejection of such practices, that are perceived as pleasurable and as markers of identity, resistance may be strong. This fits in well with US podcaster Natalie Wynn's description of the Angry Jack effect. The mere existence of the vegan at the barbecue is an existential threat and a direct challenge to the enjoyment of Jack who desires guilt free meat consumption (Wynn 2018). Freud argued that a pleasure once enjoyed can't easily be forgotten, 'whoever understands the human mind knows that hardly anything is harder' [...] 'than to give up a pleasure' once it has been experienced (Freud 2001: 145). The lifestyle element of green politics can be challenging because it has a cultural component. Cultures are both important and contested. Pleasurable practices, if we agree with Freud, may be important parts of a shared culture.

Marshall's visit to a meeting of members of the right-wing Tea Party is instructive in this regard (Marshall 2014: 17). Picked up in a seven-ton Ford Excursion, an extended-length sport utility vehicle, by his hosts, he found the Tea Party supporters were keen to demonstrate their hostility to the conventional science and policy making around climate change. He notes that they are really no different from environmental groups he is used to meeting, opinionated, passionate and strongly committed. While in his account he doesn't focus on the cultural aspects of their identity that might be challenged, he identifies that their key anxiety is around 'control'. Climate change policy, according to Tea Party members, is not even about climate but is instead an excuse for big government to reduce personal independence. Climate change is a convenient excuse for more 'control', measures like the largely forgotten Agenda 21 environmental push are seen as products of plans to restrict and dominate others. Control may involve restricting what we find pleasurable including driving, gun ownership and eating meat. Climate change is added into a framework that stresses the conspiratorial nature of left-wing politics based on an enemy who wants to interfere with every aspect of people's lives. Corporations are seen as an enemy that will destroy lives by those of us on the left: we have named an enemy. For the grassroots right, government is the enemy, a force for control in a planned society. Marshall hints at Nietzsche's dictum that when we fight monsters, we risk becoming monsters. In a polarized climate change debate the following script is perhaps used by both sides. With the end of the cold war, they, the other side, needed a new excuse, in the form of a new enemy, to justify increasing their power and control:

> So they created a story around their political worldview designed to play to people's fears and weaknesses with us as the enemy. They try to play the moral high ground but their real motives are money and political influence. They claim they are weak, but actually they are much more powerful than us because they have the support of larger funders with over political interests and because they are promoted by a lazy and biased media. (Marshall 2014: 40)

Marshall's plea to reject friend-enemy thinking can be challenged. For a start, there is a risk of ignoring power differentials. Equivalence and the rejection of the friend-enemy distinction can end up with weak and strong, the oppressed and the oppressors, as falsely equal. Yet Marshall's discussion does have the virtue of making this issue visible; we construct politics as enemy-friend contests. In turn, our political opinions are intricately linked to identity, I guess given the fragility of much of our identity, we can experience brittleness and are inclined to put emotional energy into argument. All 'of these enemy narratives seem entirely natural to the people who hold them'. (Marshall 2014: 41).

In *Living in Denial*, a study of why Norwegians often work to ignore climate change, the sociologist Kari Marie Norgaard relates knowledge to emotion and emotion to political economy (2011). In economies dependent on natural resource extraction, denial of climate change is obviously stronger. Often, the extraction of fossil fuels is found in societies where Indigenous nations, whether the Sami in Norway, First Peoples in the USA and Canada, or original peoples of the Peruvian Amazon, have been marginalised and even, on occasions, massacred. Of course, in some regions, the Indigenous were exterminated totally and cannot be remembered; even their memory has been erased, so in Indiana or Seattle city dwellers may have forgotten the history of the land they walk upon. Attempts to move beyond friend-enemy distinctions that ignore brutality and erasure of whole communities are difficult to justify. Friend-enemy distinctions may be embedded within particular social structures, products of histories that demand something more than dismissal.

Amongst an array of theorists of social knowledge, Norgaard references the work of Arlie Hochschild, whose book *The Managed Heart: Commercialization of Human Feeling,* links apparently private emotion to social structures (Hochschild 1983). It can be argued that what we 'feel' is shaped by social forces and the society we live in. The work of thoughtful climate change communication researchers like Marshall and Hoffman might be seen as missing these links between personal emotional landscape and political economy, along with social structure.

For all of the reasons discussed in this chapter, it is apparent that science does not translate automatically into belief and belief does not neatly translate into action. The more abstract, rational and nuanced the message, the less likely it is to be received, acted upon or even noticed at all. Communicating climate change science is challenging. One response is to be found in the work of Natalie Wynn, whose *ContraPoints* podcasts are entertaining and enjoyable. Her work assumes that we need to move beyond written information if we are to break down enemy-friend distinctions and persuade others. Image and sound are emotionally charged tools of vital communication. Since 2014 when Marshall wrote his book, social media has exploded with memes and podcasts, and audio visual platforms dominate political communication. While Wynn is just one amongst an army of 'progressive' podcasters, I feel she does something which is, perhaps, uniquely interesting. As well as producing entertaining material, she takes an open approach. Typically she acts pairs of characters locked into debate; this is real engagement, even if it is fictional. Her podcast *The Apocalypse* which attempts to change the mind of climate change sceptics is an ornate affair:

> Marie, a slender woman wearing white lingerie and glitter-encrusted nails, gets into her bath with a bottle of Moët and calls for her servant Antoine. When the door opens, it's not Antoine, but another woman in a lab coat and a purple wig. 'The Doctor,' as the visitor is known, has come to force Marie to watch an educational video about climate change. The pair argue, insult one another, and eventually encounter a personification of the sea, who's played as a raunchy cross between Ursula from *The Little Mermaid* and the child-devouring Cronus of Greek myth. (Mark 2019)

Natalie Wynn uses lush sets and sultry lighting to set the mood, while composer Zoë Blade provides haunting music, 'to forge a distinctive aesthetic that can be described as a kind of high-concept burlesque, drenched in neon'. The centrepiece of the ensemble is Wynn herself who plays Contra and indeed all the other characters (Mark 2019). She argues that talking to individuals we disagree with in an open way is essential if we wish to change minds,

'It's not just about calling someone out and using logic,' she explains, 'because there are emotional and psychological reasons that people hold their political convictions. From a psychological standpoint, you have to empathetically enter a person's world; not just why do they think what they think, but why do they feel what they feel? Repeat *that* back to them and you can really gain traction.' (Hall and Brownstein 2019)

In turn, she suggests that the alt right use well established emotionally charged narratives to promote a simple but compelling message:

There is something about far-right populism in particular that just grabs people on a basic, emotional level. Donald Trump's messaging and the messaging around Brexit grabs people almost at the level of primordial fear. They set up a big 'other'—a scary invasion of immigrants—and they promise protection. The symbols are so basic, like a wall. It plays pretty well in the circus of ideas. (Natalie Wynn in the *Economist* 2018)

Heidi Ravven's book *The Self Beyond Itself* indicates that while we feel we are individuals with free will who make moral choices, our opinions are often socially constructed and collective:

Ethical interventions are established at the level of description far more than we realize. Once the description has been set, the actions merely follow. So the basic fight is for the airwaves and winning the description wars – those who fund the Rush Limbaughs of the world are acutely aware of this. And it's social groups, much more than individuals, who sign on to such opinions. These deep sociological factors creating and maintaining groups underlie the more explicit group adoption of attitudes and beliefs. (Ravven 2013: 126-127)

Climate communication, despite the attempts of those like Hoffman and Marshall to make it more sophisticated, may fail for another reason. It is based on a perhaps flawed narrative of social change as a product of individual action. The individualist model is that individuals are transformed, do good things such as voting for carbon positive politicians and cutting their own emissions. Communication to allow intervention to shift an entire social system, what I describe as a mode of production, is a different project. Even if we argue that such fundamental transformation is unnecessary, would take too long or is unrealistic for other reasons, can we trust politicians to introduce appropriate policies? Chapter Nine discusses the extent to which politicians are part of a state machinery which, far from working for the common good, may be seen as a flawed and repressive institution.

Chapter Nine
Don't Trust the State

Climate change [acts] *as justification* for the commodification of the atmosphere and, especially, for the commodification of the gas, carbon dioxide. In this frame, climate change is viewed as the latest rationale for converting a public commons into a privatised asset – in this case, the global atmosphere. 'Ownership rights' to emit carbon dioxide are allocated or auctioned between entities, alongside the attendant machinery of the market which prices and regulates the commodity. (Hulme 2009: xxvii).

Much of the debate over climate change works with the following model: Campaigners rouse public opinion, the public votes for politicians concerned with climate change, the politicians win elections and introduce policies to reduce emissions. Problem solved. Having discussed some of the challenges with communicating climate change, with climate policies, and the strengths and weaknesses of Green Parties, trade unions and social movements, it is essential to also consider the nature of the state. Within the broad left the solution is often the state. Debating the state, and looking at the problems of a state solution, might be seen as something practiced by those on the right acting in bad faith to discourage action on climate change. While I am not an anarchist or a libertarian, I must admit I am sceptical when it comes to the state. There are plenty of reasons why states can't be trusted to smoothly legislate for a better world. Like nearly everything else around climate change, things are a little more complicated. While advancing towards serious work to slow and eventually halt climate change emissions, as well as adapting to its growing effects, we need to think with some care about the politics appropriate to doing so. We need to be aware of what can go wrong with statist approaches to climate change and indeed statist approaches in general.

On the free market right, anti-statism has long been a way of challenging the left, but became especially important during the 1920s and 1930s. After the October 1917 Russian Revolution put the Bolsheviks in power, Marxism was rapidly gaining attention and status. Across much of Europe, reformist social democratic parties seeking to represent the working class were also growing fast. A number of sceptics, often Austrian, moved from advocating socialism to pointing out the likely failure of any socialism based on the state. Friedrich von Hayek, Joseph Schumpeter and others created the Austrian school of economics, representing a bracing free market approach. In the US, the Chicago School, famous for its right-wing neoliberalism, was founded by Frank Knight, who was originally a socialist, but swung back briefly to the left calling for a vote for the US Communist Party in the 1930s. James Buchanan, another US economist on the right, again with Austrian sympathies, also started as a socialist (Burgin 2009).

They all argued that central planning undertaken by the state led to chaos. Hayek (1937) was representative in asserting that this was due to 'the knowledge problem': millions of goods and services could not be allocated efficiently between millions of consumers in a centralised way. States fundamentally cannot cope with the complexity of economic decision making. But if the battle over the political economy of planning, used to indicate that socialism was bankrupt by the right, was significant, it is also the case that there is a strong strain of anti-statism on the left. Marx and Engels famously called for the 'withering away of the state' (Surin 1990). Lenin noted that the state was not a neutral instrument but the product of the '*irreconcilability* of class antagonisms', and had to be smashed (Lenin 1965: 7). While I am not fundamentally against the state, I do feel the apparent and often unquestioned model of social change based on consciousness raising, election victories and legislation can lead to problematic results.

Whose common future?

Perhaps what it comes down to is power and knowledge. We assume that states have power and, unless corrupted, use such power in a benevolent way. We also assume that the state has knowledge; it can provide solutions and put them into practice. The state is assumed to be both an institution that works efficiently as an instrument and can be used by whomever controls it. The Marxist critique is one of power: the state is not neutral, but works for a ruling class. The market-based critique is one of knowledge: the state lacks intelligence, it makes a mess of things.

One criticism, developed largely on the right but worth consideration, is the problem of unintended consequences. When we do 'X' to achieve 'Y' we may be surprised to later find that 'Z' occurs. The free market right argue, for example, that a minimum wage, rather than raising the living standards of the poorest, instead makes them even poorer because it produces unemployment. Likewise, they suggest that rent controls, instead of making rents cheaper and helping people to afford accommodation, produces shortages. The belief in the market can become a tired dogma, markets fail and tend in particular to increase inequality. I think too that on the minimum wage and rent controls, such consequentialist arguments are wrong. Specifically, while raising the wage rate in market theory tends to increase job losses, in practice it often makes no difference (Ritholtz 2019). Discouraging individuals from renting their property, by lowering rents, logically reduces the number of rentals, however it may encourage property to be sold, increasing the supply of homes and reducing accommodation shortages (Walker and Jeraj 2016). The risk of unintended consequences should however be acknowledged. Every policy, however attractive in theory, may have some kind of negative consequence in practice.

The knowledge problem is likely to produce unintended negative consequences because policies which are applied nationally or internationally are likely to disregard local conditions. This has already been a feature of the existing global framework based on 'cap and trade'.* Significantly, this is a policy that uses market forces, however the unintended consequences of the market are often ignored by right-wing critics of government action. Governments have been able to offset their countries' contributions to climate change by sponsoring projects internationally. This has led to corruption, displacement of Indigenous people and various other sharply destructive effects. The system has been commonly compared to papal indulgences. These occurred when medieval Popes allowed wealthy sinners to pay a cash or property contribution to the papacy to guarantee a place in heaven after they died (Smith 2009).

To cite one example, the Kenyan government forcibly removed the Sengwer people from the Embobut Forest. The assumption was that the Sengwer were 'squatters' destroying the forest, and had to be removed. After a Sengwer man was killed by the authorities, the EU suspended funding for the Kenya Water Towers Protection and Climate Change Mitigation and Adaptation Programme (Amnesty International 2018). There are many other examples of where schemes to combat climate change incorporate assaults on Indigenous people. REDD (reducing emissions from deforestation and degradation) schemes allow governments to conserve forests in exchange for global carbon credits. Often these lead to land grabs, displacement of peoples, and may do little to reduce emissions or protect carbon sinks (Leach and Scones 2015).

Corruption is not a unique feature of renewable energy, but it certainly is not immune from it either. In the north of Ireland, a peace process has been put under extreme strain by a renewable energy scandal. The Northern Ireland Executive

---

* For example, https://foe.org/california-legislature-just-extended-cap-and-trade/

and Assembly had, despite some tension, seen a coalition between Sinn Féin and their previously sworn enemies the Democratic Unionist Party (DUP). This structure ended a conflict that since the 1970s had led to the loss of thousands of lives. Martin McGuinness, a former butcher's boy turned Irish Republican Army leader turned Sinn Féin politician, resigned as deputy first minister because of a renewable energy corruption scandal. He stood down after ten years in office in January 2017 because of the so called 'Cash for Ash' affair. The Renewable Heat Initiative (RHI) aimed to encourage farmers and others to install renewable energy, including wood burners, to reduce fossil fuel emissions. The DUP's Arlene Foster allegedly failed to monitor this adequately, with Ulster farmers, a key group of her party supporters, apparently exploiting the RHI. The scheme risked promoting moral hazard with participants picking up large grants to heat their homes and farms, leading to an overspend of £490 million (McBride 2019). The critical error in the scheme was that the subsidy on offer was greater than the cost of the fuel for the biomass boilers that claimants were using. That effectively meant there was an incentive for people to burn more and more fuel to earn more and more money. A news report noted that within days of the RHI scheme being launched, renewable energy firms were marketing their boilers with the line that buyers would receive '20 years of free heat'. Such was the frenzy to sign up that one individual 'was willing to pay have a boiler flown from Austria to County Fermanagh at short notice to beat the deadline'. (McDowell 2018)

Corruption is a feature of much activity, subsidies for coal and oil might be seen as innately corrupt, however the cash for ash scandal is instructive. Support for renewables can be exploited for inappropriate or even criminal gain. Of course, any adverse consequences of climate change policy will be weaponised by being taken out of context and married up with a scary meme by defenders of a fossil fuel status quo. This, I think, makes it more, not less, important to be aware of unforeseen consequences and to promote the understanding that policy changes are not magical solutions with purely positive effects.

Such abuses suggest not only unintended consequences, but institutional bias. That the state, it is assumed, governs for all and considers a plurality of interests, seems to be the logic behind Green Party electoral strategies and the state orientated demands of Extinction Rebellion. Of course, there is an understanding that corporations and the powerful may have more influence, but this is seen as a corruption of the function of the state and one that can be overcome. It might be more realistic to argue that states are born out of violence and act on behalf of the few, not the many. The state is a complex piece of interlocking machinery, which works not to maintain and advance the common good but instead to manufacture obedience. The lesson of history told over many chapters is that states have been more effective at domesticating their opponents than of being transformed by such radicals.

The state tends to take demands but reshapes them in ways which maintain the power and wealth of those with existing power and wealth. Policing acts to repress movements, while classic techniques of incorporation reward the most moderate voices. Moderation is primarily a euphemism for maintaining the interests of existing elites and their arrangements. Policy suggestions can be delayed and put off with various commissions and investigations, until attention has dropped and policy work can be forgotten. Political parties tend to function in an oligarchical way, often with relatively little input from members. Electioneering is an expensive business, and in many countries billionaires are cutting out the middle people and ruling directly.

Even in democratic states with strong constitutions to protect individual rights, repression is common. In Britain, not of course a country with a written protective constitution, environmental activists were spied on by the state during the 1990s, despite their entirely non-violent stance. The Spy Cops scandal revealed that some women activists had even had children with men who turned out to be undercover police

(Lewis and Evans 2013). In the USA, police racism remains common with African Americans innocent of any crime being shot. First Nations protests against oil pipelines have also been put down with extreme violence in the USA (Estes 2019). Rapid growth in technology platforms suggest we live in an increasingly surveillance-based society, where it becomes easier to police dissent (Greenfield 2017). One can imagine repression too in the defence of climate change policy, with the removal of Indigenous communities to make way for renewable energy schemes.

The anthropologist James C. Scott, in his major work *Seeing like a State*, provides a largely epistemological critique of statism. Epistemology is the study of knowledge and Scott's argument, like that of Hayek and other Austrian economists, focuses on knowledge. However, he rejects their devotion to market-based alternatives and their apparent silence over issues of power and oppression. Scott argues that the advances of science contributed to an epistemological arrogance, common to many on both the left and the right during the early twentieth century:

> It is best conceived as a strong, one might even say muscle-bound, version of the self-confidence about scientific and technical progress, the expansion of production, the growing satisfaction of human needs, the mastery of nature (including human nature), and, above all, the rational design of social order commensurate with the scientific understanding of natural laws. It originated, of course, in the West, [...] as a by-product of unprecedented progress in science and industry. (Scott 1998: 4)

He believed that this gave rise to an uncritical approach to science, producing an epistemology that meant that everything was potentially knowable became an unrealistic dogma. The notion, which he describes as 'high modernism', assumed that scientific knowledge could be used to plan a better society (Scott 1998: 4). Trust in science was unsurprising in the early years of the twentieth century, as a range of sciences were advancing fast. Scott argued that such a high modernist approach led to a desire for legibility, so as to make the invisible and confused visible and clear. He noted that nomadic peoples tend to create particular unease for state projects and there were constant, largely unsuccessful, attempts to settle Roma and similar groups. The state project tends to involve naming citizens, recording them by census, and providing identity cards or official insurance numbers. Much of what we take for granted about a state like France or Britain is based on an evolution over many centuries, to name, to make clear and to control populations. States in their modern and modernist form are the product of particular processes and involve a paternalistic approach. Of course it is easy to see how a kind of benevolent carbon paternalism could be created, monitoring each citizen's use of energy. Perhaps the concerns of various climate contrarians have a rational basis in this regard if in no other. Scott argues that legibility provides the possibility of large-scale social engineering, the modernist ideology inspires a desire for such social engineering, and an authoritarian state, where it exists, provides particularly strong tools to complete the job.

Scott didn't study climate change but looks at various high modernist episodes of disaster. His point is that knowledge of a society is always incomplete, the modernist assumption that it is possible to promote various utopian schemes is an illusion and disaster often strikes. Of course one might note that if the modernist project was so easily realisable, if states really were competent and powerful, we might witness ever more effectively managed forms of benevolent or hostile authoritarianism.

Scott, incidentally, engages in a detailed and lengthy critique of Lenin's revolutionary strategy which he sees as a scientific and high modernist enterprise, but notes:

> By focusing on Lenin's high-modernist side, we risk simplifying an exceptionally complex thinker whose ideas and actions were rich with crosscurrents. During the revolution

he was capable of encouraging the communal seizure of land, autonomous action, and the desire of rural Soviets 'to learn from their own mistakes'. (Scott 1998: 167).

I would suggest that while the period that Lenin lived in was marked by modernism, Lenin was a creative, flexible and dialectical thinker (Harding 1980). Yet while Lenin died only a few years after the Bolshevik Revolution, and these were years marked by civil war and some confusion, a modernist perspective was common to twentieth-century Marxist approaches to governance. Respect for technological development, centralised control, and rejection of diversity were all part of a modernist Marxist project. Indeed they were common to the reformist left, including thinkers such as the Fabians, and as Scott notes, commonly held across both right and left. The assumption that the environmental disasters of the Soviet Union, such as the draining of the Aral Sea, were a product of Stalinism, is only a partial explanation (Pryde 1991). For all the supposedly unique tyrannies of Stalin, Trotsky, along with Stalin and many others on the left and indeed the right, held a modernist technocratic vision based on large-scale engineering. The assumptions behind high modernism were common to groups as diverse as anarchists and imperialists in this period. While creating some benefits, modernism also created many disasters.

While we live in a more sceptical age, there is clearly a danger that large-scale attempts to deal with climate change could lead to unforeseen consequences and various environmental and social problems. The possible use of geoengineering is perhaps the most worrying example (Hulme 2014). A Green New Deal, while essential, might lead to unforeseen negative effects and perhaps more predictable power grabs by elites. The political ecologist Giacomo D'Alisa has criticised 'emergenciocracy' as potentially dangerous and counterproductive, broadly for the same reasons (D'Alisa 2019). This is not to argue for doing nothing, nor to promote the denialists who might pick on a possible consequence of cleaner energy as a justification for maintaining a destructive fossil fuel-based society. Critiques of the state can be mobilised, indeed are often mobilised, to promote market economics. However as discussed, markets tend to generate inequality, ecological damage and a host of other problems. Markets are no solution to the problems of climate change and other environmental ills. Yet it may be possible to move beyond the market and state binary, to develop more creative, diverse and less oppressive forms of governance.

Interestingly, despite the often poor environmental record of the Soviet Union, the country seen as closest to achieving an environmentally sustainable path is Cuba. During the special period in the 1990s, when cheap oil from the Soviet Union ceased to flow, attempts were made to move Cuba to a low carbon economy, with some success. Public transport and low carbon permaculture to deliver food supplies along with an expansion of renewable energy, mean that by some measures Cuba tops sustainability indices (Trinder 2020).

There are other signs of hope. Invisible to much of the world's media and public opinion, in pockets around the world, a strange but productive encounter is taking place. This encounter is already producing action to promote ecology, and directly addresses the concerns of state pathology noted here. This is an encounter, to simplify the narrative a little, between Lenin and the anarchists. Rojava in Northern Syria is one example.

The Rojava Revolution

In Northern Syria, the Kurds and their allies have a republic. A form of governance without government, based on the concept of democratic confederalism derived from the thinking of the New York anarchist Murray Bookchin. In the wake of the Arab Spring, there was an uprising against President Assad's Baathist regime. A civil war ensued and the conflict eventually became centred on the so-called Islamic State. In the chaos, the Kurdish minority in the north of the country took power in the Rojava revolution, Rojava being a Kurdish term for the west. The politics of Rojava are still unfolding. As I write, there are fears that Turkey, which has invaded Syria, will wipe out Rojava. But while this is neither the place nor is this the right time to draw up a full account of Rojava, its existence provides an account of how governance without a traditional state may be possible (Knapp *et al* 2016).

The People's Protection Units (YPG) are a citizens' militia, the defensive force in Rojava. They have fought and defeated the so-called Islamic State, at considerable cost, with many thousands of people killed (Knapp *et al* 2016). They are informed by the writings of the historic Kurdish leader Abdullah Öcalan. As I write, Öcalan remains in prison in Turkey. The Kurdistan Workers' Party (PKK) were originally inspired by a Marxist-Leninist approach, taking heart from the success of the Viet Cong in defeating first French and then US forces in Vietnam. In recent years, ideological transformation has been undertaken, suggesting a need to create a non-state, decentralised, diverse and feminist form of governance. Bookchin's view was that while states were oppressive, forms of local and participatory democracy could be liberating. His democratic confederalism drew upon the examples of the city states of late medieval Europe and the assemblies in New England, where the community made decisions at open public township meetings (Biehl 2015). Such democratic confederalism has been introduced in Rojava by the Kurds and their allies. In the Chiapas of southern Mexico, the Zapatistas have innovated their science of revolution to create a similar non-state form of governance, drawing on Indigenous traditions of self-governance. Like the revolutionary Kurds, their roots are also to be found in Marxist-Leninist ideas, as well as traditional forms of self-organisation, liberation theology and anarchism (Oikonomakis 2019).

Self-governance, with a focus on commons and collective ownership systems have been researched and theorised by the Nobel winning economist Elinor Ostrom (1990). She argues that self-government is the norm in all societies, in that we are constantly negotiating with others and making collective decisions. For her, politics runs through all aspects of life and is not just a product of the state. She can be seen as advocating and developing a science of association, showing how trust and cooperation could be built. We can constantly push for the decentralisation of power and investigate ways of making diverse local democratic systems work; Ostrom's work acts as a thoughtful guide to doing so (Wall 2017). She applied her methods to show how climate change could be tackled, both via adaption and reducing emissions, by adopting diverse democratic participation (Ostrom 2012).

Given the timescales needed to deal with climate change, an opposition to the state might be seen as far more unrealistic than a mere demand to replace capitalism. However, we should be aware that the state isn't our friend, and where it does positive work, this is perhaps an exception. We should not trust the state but instead treat it with some caution. Even if the state was both wise and just, transforming a way of life, a culture, what I would describe using Marx's term as a mode of production, cannot be achieved purely by substituting one group of politicians with another. Deeper change is necessary. Fundamental transformation, the subject of my concluding chapter, is vital.

Chapter Ten

Green Lenin, Green Machiavelli

> I compare fortune to one of those dangerous rivers that, when they become enraged, flood the plains, destroy trees and buildings, move earth from one place and deposit it in another. Everyone flees before it, everyone gives way to its thrust, without being able to halt it in any way. But this does not mean that, when the river is not in flood, men (*sic*) are unable to take precautions, by means of dykes and dams, so that when it rises next time, it will either not overflow its banks or, if it does, its force will not be so uncontrollable. (Machiavelli 2019: 83)

So what do we actually do? How can we act to slow and reverse the rising emissions that are driving climate change? The outlook is pessimistic, a pessimism that if anything I have deepened over the last nine chapters. The science is alarming, climate change is clearly with us now, is having sharply negative effects, and emissions continue to rise. Hardly a day goes by without more evidence for catastrophe. The subject of this book is a conversation that we needed to have had fifty years ago, followed by swift action. Given that we needed to act yesterday and didn't, delay is even less acceptable now. Yet climate change isn't a quick fix. My argument, which is hardly original, is that it is a crisis embedded deeply in the dynamics of our present global society. The time needed to change a whole civilisation, an economic and social system, is not something easily measured in years or even decades. The practical politics of climate change creates a temporal anomaly; we need to act quickly, but the solution requires radical and systemic change that would take many, many years of work.

The model of conveyor belt radicalism is also broken. By this, I mean the model of social change accepted by some of us, as a kind of common sense, is unlikely to succeed. Conveyor belt radicalism can be conceived in the follow way: 1) demonstrations and vibrant protest; 2) inflame popular opinion; 3) this is used to inspire politicians to action; 4) policies to tackle the climate crisis are introduced; 5) a happy stable climate is the result. The model has always been flawed. The crisis cannot be solved by easy reforms, it is systematic. Moreover, as discussed in the last chapter, states are imperfect bodies and must be treated with caution.

Neither does human action rationally and easily occur. We are part of nature too, and work with particular biases and qualities. While the science is rarely static, it is possible to attempt to examine human motivation scientifically. Investigations from social psychologists and others suggest that we humans have built-in cognitive biases; we use frames to understand the massive quantities of information thrown at us every day. We tend to bend reality into narratives that we are familiar with, oversimplifying and distorting what is out there. Indeed, as previously noted, there is now a well-entrenched narrative of denial which has absorbed significant segments of the right, especially in countries like Australia and the USA. It seems impossible at the moment to think of a Republican President in the US accepting climate change as a significant problem. This isn't a mask; it is now the face. The denialist case, debunked thoroughly, exists in a zombie form. However, operating on the zombie's brain does little to kill the denial. Every particular point, from 'temperatures aren't rising', to 'they are rising but this is caused by solar flares', to 'rising temperatures are beneficial', has been successfully challenged. Nonetheless the denial is, if anything, stronger on the right than ever. We are dealing with deep seated identity politics. To many on the right, climate change denial is a component of an articulated set of beliefs. So whatever the protest, a solid and influential section of society is likely not merely to ignore climate change, but to actively work against solutions.

It is not just the existence of denialist right-wing politicians that disrupts the flow of the conveyor belt model of radicalism. An additional challenge, given that political communication is based on narratives rather than rational choice, is the issue of attention. We live in an economy of attention with thousands of images and sounds projected at us digitally. Businesses, political parties, social movements, and others try to get our attention for a particular message (Doran 2017; Lanham 2006). Climate change concern has been influenced positively by social movement action. As I write, XR has been gaining huge attention through disruptive direct action. Maybe this means the belt isn't broken at an early stage. At least in Britain, massive XR actions with their spectacle including pink boats, mass arrests, attention-grabbing and semiotically coherent strategies, worked some magic, encouraging politicians to declare support for a climate emergency.

The near certainty is that attention will wane because it is difficult to keep attention on one particular issue. Once attention has fallen, the assumption is that business as usual, in its environmentally destructive form, will continue. If Heathrow Airport Holdings Limited are patient and wait a little they will be rewarded with a new runway at Heathrow. If universities delay, they can maintain their investments in oil once protest fervour has fallen away. Companies like BP and Shell can wait it out and within a short few months ministers will no doubt be celebrating the discovery of new oil sources and offering tax breaks to get the black stuff gushing out of the North Sea.

So far so bad? So what do we do? For a start, we need to act now to halt or, at least slow, emissions. At the same time, we need to work towards creating a post-capitalist ecosocialist society. Integrating these tasks is possible. We need to deepen the sophistication of our understanding of the crisis and focus upon the many vexed questions of strategy. We must, in turn, translate strategic thinking into focused action.

There are some actions that we can take immediately to slow the rise of emissions. Directly challenging companies producing fossil fuels is perhaps the most obvious and useful. Institutions including religious organisations, trade unions, local authorities and their pension funds, have huge amounts of money invested in such firms. Divestment campaigns to remove such funds are vitally important. Before helping to launch XR, co-founder Roger Hallam was a doctoral student at King's College in the University of London. He found that King's had invested in oil companies, so launched a vigorous campaign to force them to divest. A five-week campaign involved a hunger strike and property damage in the form of spraying the slogan 'Divest from oil and gas' on college walls. The college eventually agreed to divest and to try to become carbon neutral. His militancy on this occasion prevented delay and achieved the goal. Later on trial for £7000 worth of criminal damage, Roger Hallam and an associate were acquitted (Cockburn 2019).

Oil companies engage in widespread sponsorship activity. This may be inspired by genuine concern for culture, but it helps soften their image, which in turn reduces resistance and increases their apparent moral legitimacy. Challenging this is another necessary form of action we can take now. British Petroleum's sponsorship of Shakespeare's plays has been targeted by the 'BP or not BP' campaign. Their first guerrilla performance, as the Reclaim Shakespeare Company, took part on 23 April 2012, which is Shakespeare's birthday. Two guerrilla performers jumped on stage at a Royal Shakespeare Company performance of *The Tempest*, Stratford-upon-Avon, to challenge financial sponsorship from the oil company. This was shortly after BP's Deepwater Horizon drilling disaster and the oil company's decision to start extracting highly polluting and destructive tar sands in Canada,

> What country, friends, is this? Where the words of our most prized poet / Can be bought to beautify a patron / So unnatural as British Petroleum? / They, who have incensed the seas and shores / From a dark deepwater horizon /Let us break their staff that would bewitch us! / Out damned logo! (BP or not BP 2015)

After performing this verse, the performers ripped the BP logo from the theatre programme,

> We have performed without permission at the Royal Shakespeare Theatre, the British Museum, the Edinburgh International Festival, the National Gallery, Cadogan Hall, the Royal Opera House, the Science Museum, the Roundhouse, the Noel Coward Theatre, and in Tate Britain. (BP or not BP 2015)

We can also vigorously oppose new forms of fossil fuel extraction. Campaigns to prevent drilling for more oil, opening up open cast coal mines and fracking for gas can and should be a focus for ecological militancy. Focused solidarity with Indigenous people is also vitally important. In Brazil, the Indigenous are strong defenders of the Amazon and as such are the targets of the beef barons, soya kings and their political supporters such as Jair Bolsonaro. Indigenous people are often forgotten in serious discussions of climate action. The battles to defend the rainforests need to be joined. In turn, struggles to defeat extraction and defend the forests will be attacked by corporations and their powerful allies. Practical support for environmental prisoners and fundraising for legal campaigns is another vital dimension of what we must do if we are to be effective in challenging rising temperatures.

Another approach is to argue that such practical action to tackle the climate crisis must rest on new psychological, spiritual or philosophical principles (McIntosh 2008). Perhaps we need a deep ecology that respects other species as a foundation for what we do? Eroding the binary between humanity and the rest of creation may be viewed as important from this perspective. Humans are not the measure of all things, and it is not only humans that have the potential to act. There are plenty of anecdotes about other species acting with agency and intent. Typically I am charmed by a news story of a bear and a wardrobe. In June 2019, a black bear in Butler Creek, Missouri, managed to open an unlocked door to a house. Bolting the door from the inside and becoming trapped, in its frustration it smashed up furniture before taking a nap in a wardrobe (Anon 2019d). I am always emotionally captured by animals with apparent agency. Yet for the philosopher and sociologist Bruno Latour, perhaps the real hero of the story is the wardrobe, or he might muse on the tragic damage inflicted on the room's other furniture by the bear. Latour's striking, and at first sight bizarre, approach to metaphysics is the main target of Andreas Malm's book, entitled *The Progress of This Storm*, discussing the philosophical foundations of effective action on the climate crisis. While Malm (2018) covers other thinkers, the last chapter contains an extensive discussion of the environmental historian Jason Moore, most of the book focuses upon a critique of Latour. Indeed criticism of Moore is based largely on the supposed similarities between Moore's analysis and that of Latour.

Graham Harman, a pioneer of object-orientated ontology, argues that Latour's core insight is that,

> No actor, however trivial, will be dismissed as mere noise in comparison with its essence, its context, its physical body, or its conditions of possibility. Everything will be absolutely concrete. (Harman 2009: 13)

According to Harman, Bruno Latour's epiphany came in 1972. Stopping on the road from Dijon, he was suddenly enlightened with the following notion,

Nothing can be reduced to anything else, nothing can be deduced from anything else, everything must be allied to everything else. [...] This was like an exorcism that defeated demons one by one. (Latour in Harman 2009: 13)

This is an almost magical vision with much to commend it. Rather than humanity being the measure of all things, all things in contrast have worth. In turn, all things can be seen as actors, even if they lack intentionality in the strict sense. Any actor acts to shape and influence. It is easy to ridicule this: pigeons, emperors, rocks, pieces of wire and even ideas are apparently actors for Latour. Yet Latour's insight is philosophically challenging and important. Where do we draw the line between actors and non-actors, simply between humans and the rest of creation, animal, vegetable, mineral and beyond? How will we deal with the technological singularity when artificial intelligence becomes more intelligent than homo sapiens (Creighton 2018)? How should we react to the report from the *Economist* that capuchin monkeys' tool use has dramatically evolved in sophistication over a period of 3,000 years and continues to do so (*Economist* 2019b)? While the capuchins' tools might not have agency, Latour nonetheless addresses a fascinating and difficult debate as to where we draw the boundary, or whether to abandon drawing boundaries, between the natural and the social, between actors and the acted upon.

Yet the sweeping ontological claim of a kind of democracy of things put forward by Latour is too general, Malm argues, to advance a precise analysis of climate change and in particular the role of capitalism in promoting it. Malm defends Marxist historical materialism as a means of analysis that values both the natural and the social. He provides a readable summary of Marxist attempts to theorise ecological crisis, from Marx and Engels' work, via Bloch and Marcuse, through to the insights of John Bellamy Foster. Foster argues that Marx in *Capital* uncovers a break in the metabolism between nature and human society, which creates environmental problems. Foster and a number of other researchers have used this concept of an ecological rift to analyse current issues including the climate crisis (Foster *et al* 2011). There is a rich tradition of ecological thought within Marxism; Malm argues that while understanding that humans are part of nature, Marxism suggests that the social and the natural can best be understood by making these two categories analytically distinct. Malm stresses that his task is political, noting the 'purpose of such analysis, then, should be to feed into resistance or, preferably, revolutionary ecological practice'. (175). In dealing with climate change I would agree with Malm that we need 'Less of Latour, more of Lenin' (2018: 98). Lenin argued, correctly, 'without revolutionary theory there can be no revolutionary movement' (Lenin 1961: 369).

Lenin's phrase has been used to create closure and certainty. Instead it should be a way of opening debate that leads to diverse and potentially effective pathways to change. Such change works in ways that are both revolutionary and immediately practical. Closure is, of course, more familiar. For example, the apparently naïve environmentalists are called out by the Marxists for their naïve misconceptions; green failure is contrasted with red promise. The examples are too numerous to mention, but one which struck me as instructive and typical was a Facebook post from a local Marxist in my area about the Irish Green Party. As noted in Chapter Four, Greens have often been criticised by the left, and the Irish Greens' coalition with the centre right party Fianna Fáil led to disaster. Fast forward to 2019, the Irish Greens have recovered, winning numerous local council seats and electing a number of Members of the European Parliament. They have taken a decision not to rule out a new coalition with either Fianna Fáil or the other dominant Irish Party, the equally centre-right Fine Gael, to the dismay of many. The Facebook post noted that this was a product of their failure to embrace an anti-capitalist perspective, and lauded a broadly Leninist approach. I think this is too simple; I certainly agree that an anti-capitalist perspective is necessary, but an anti-capitalist approach is not sufficient to be politically effective. Lenin's dictum should inspire us to think about theory, so as to inform specific practices. Let us not use it as a flag to wave, indicating political superiority to others. Theory should, in my view, work to provide tools for social change. Often it can instead act more like a political symbol, which is used prematurely to end discussion and invoke a narrow party loyalty.

Lenin, whose theory is complex, contradictory, and debated in many ways, would no doubt have thought differently about the situation today than he did of Russia in 1917. I am not proposing that we simply read everything he said and use his concepts as a template. Often, reference to his work is a mark of supposed revolutionary fidelity rather than a guide to action. Lenin's bust in marble, a stack of his collected works behind glass doors, or a newspaper headline with his image, can be forms of symbolic currency. Theory can also be an academic pursuit which is unrelated to any form of material practice. Theory can be a means of silencing others and policing membership of an organisation.

Theory can appear to be an esoteric distraction. Mackenzie Wark the Australian born critical thinker observes:

> Let's invent new metaphors! Personally, I like the #misanthropocene, but don't expect it to catch on. Jason Moore prefers the *Capitalocene*. Jussi Parikka the *Anthrobscene*. Kate Raworth suggests *Manthropocene*, given the gender make-up of the Anthropocene Working Group considering it as a name for a geological era. Donna Haraway offers to name it the *Chthulucene*, a more chthonic version of Cthulhu, the octopoid monster of H. P. Lovecraft's weird stories. (Wark 2015: 223)

What I am suggesting is that we use available theory to examine aspects of the practical politics of climate change to try to think about what we do, with more care. 'Theory' is slightly funereal and discouraging word; what I am getting at is the idea that we should find out more about any research that helps us understand how something works. With better understanding, perhaps we can be more effective. Concepts can be used to help change the world. I guess some of Lenin's theory is still directly useful in regard to climate change politics. For example, although the world has changed since World War I, the theory of imperialism which he helped develop help to explain the origins of the war and might still explain much of the origins of climate change. This is, of course, part of the argument made by Andreas Malm in *Fossil Capital*. I don't think, however, that Green Parties or social movements like XR can overcome all their various weaknesses by asking their members to sit down and work their way through *The State and Revolution*, *What is to be Done?* or even volume 38 of Lenin's *Collected Works*, which includes his all-important notes on Hegel. Yet we can imagine Lenin writing *Extinction Rebellion: An Infantile Disorder?*

Lenin attempted to provide a conceptual understanding of how to advance revolutionary change in a particular context. In the century since 1917, neither the revolutionary nor the reformist left have achieved unambiguous success in either reforming or transcending capitalism. This is another reason why calling out either Green Parties or environmental social movements on the left is inadequate. Can we develop a theory of revolutionary transformation that promises to change everything so we can protect our planet from the ravages of ecocidal capitalism? Is revolution in the sense of profound transformation possible?

In some parts of the world, insurrection has had some success. I noted in the last chapter the existence of militant ecological communities that have risen up and achieved local forms of revolution such as the Kurds and their allies in Rojava and the Zapatistas in Mexico. These are not models that can be followed universally, and if revolution in one country is rejected by some as inadequate, how much weaker is revolution in a couple of small regions of the world?

Often, far left parties talk revolution but indulge in fairly ritualised activities of selling newspapers, preparing websites, engaging in trade union solidarity, reading theory and denouncing others on the left. As we have seen, the alternative of promoting purely electoral politics is also inadequate. The state can't easily be picked up and used as an instrument. While I am not discounting electoral politics or revolutionary insurrection in relevant contexts, I think there is another strategy on the left that might have some utility. By this I mean that it is useful now in combating climate change, and also useful as a possible contribution to the deep long-term systemic change which is so necessary. This is the concept of base-building, which has emerged amongst groups to the left of the left in the USA.

While Trump has been rallying and mainstreaming racists, climate accelerationists and religious fundamentalist homophobes, the left in the USA has also been enjoying some growth. Dominated by a two-party system, the Republicans and Democrats have tended to squeeze out even the mild social democratic left found in many other parts of the world. The USA, unlike Australia, Aotearoa New Zealand or Britain, has no electorally successfully Labour Party. However the socialist (more social democrat) Senator from Vermont Bernie Sanders was locked into a nomination battle with Hilary Clinton as the Democratic Presidential nominee in 2016. His success also contributed to the growth of the Democratic Socialists of America (DSA). Back in the 1960s and 1970s the new communist movement emerged in the USA, inspired by the Vietnam war and the revolutions in China and Cuba. It had some success in mobilising African Americans and overlapped with the vibrant and iconic Black Panther Party. It was composed of an archipelago of Leninist organisations working towards a new American Revolution. Despite its initial energy and growth, the US American right gained strength in the 1980s and onwards. Max Elbaum, a participant in the new communist movement, described its collapse into sectarianism and sterility in his encyclopaedic book *Revolution in the Air* (Elbaum 2018).

The insurrectionary Leninism of the new communist movement came to essentially nothing. Despite the growth of an electoral left in recent years, progress towards a socialist or ecosocialist USA looks slow. Even mainstream Democratic Party politics, with an emphasis on mild climate change action, looks challenged in the USA. The electoral path appears unusually difficult for the left, as the party system is restricted effectively to two parties. There is no form of electoral proportionality in national contests, meaning that a minority party will generally fail to gain any elected members. The US system is marked by a series of constitutional measures that give an advantage to the Republican Party; for example, because of the nature of state representation, Donald Trump polled fewer votes than Hilary Clinton but was still elected President in 2016. The influence of money is massive, meaning that US presidential challenges are becoming increasingly restricted to billionaires. The US is perhaps an extreme case, but both reform and revolution look like flawed strategies. The US system has some built-in political diversity, of course, with states having a great deal of power and localities often having considerable autonomy. Nonetheless it is a difficult society within which to advance left or ecosocialist politics. Work within the Democrats and you may well be absorbed, set up an independent party like the Greens and you will have far less success than Greens in most countries. Establish a party that aspires to revolution, and you are likely to be brutally suppressed like the Black Panther Party (Churchill and Vander Wall 2002). Those aspiring to revolution are often so unsuccessful that state repression is unnecessary.

The original Leninist model has, in many ways, run out of energy, but an element within it which may remain useful is the creation of dual power, which involves building alternative institutions to rival official state power. Base-building is a tactic that draws upon Lenin's concept of dual power. A base-building organisation might seek to create local trade unions, tenants' associations, legal support for prisoners, and advice centres. Across the USA the Marxist Center, a federation of militant but diverse communist groups, practices base-building. Marxist Center groups, rather than adhering to a particular ideological tradition, instead share a commitment to practical work, creating community institutions which effect material change. This stress on practical action is seen as a way of promoting an alternative to both sectarianism and protest culture on the revolutionary left (McQueeney and Parkinson 2019). A report on the founding meeting of the Marxist Center described their work in the following way,

> Base-building, simply put, is organizing the working class into institutions that are vehicles of collective struggle. This can mean challenging the rule of capitalists through industrial or tenants' unions, or it can include things like mutual aid associations and cooperatives. [the aim] is to build a 'dual power' to the capitalist state, creating a workers' society of mass organizations that are independent of any capitalist political party. (McQueeney and Parkinson 2019)

One of the strongest component organisations within the Marxist Center are the Philly Socialists. They describe their activities in the following way:

> So, what does it mean for us to be a 'base-building' socialist organization? Is it that different from being in any other socialist group? Good question, ourselves! To give you a flavor of what being in PS is like, here's a partial list of what we've been doing the past 2 weeks.

After listing various activities they note:

> This is the work we believe socialists need to commit to: rebuilding working-class institutions, going to the people directly, struggling with our enemies, educating ourselves, and taking care of each other while we do it. It's hard work, but if we don't do it, who will?? (Philly Socialists 2018)

They run a tenants' group, the Cesar Andreu Iglesias Community Garden (named after a Puerto Rican revolutionary), and a variety of other projects. Base-building in the USA isn't restricted to the Marxist Center. Others who practice it include the Refoundation Caucus, the Communist Caucus of the Democratic Socialists of America, some branches of the Party for Socialism & Liberation, the Organization for a Free Society, the Socialist Party USA, the Malcolm X Center, and Cooperation Jackson (Cooper 2018).

Base-building is closely related to the notion of organisational materialism. This is the idea that what matters is the practical results of an organisation's activism, rather than where an organisation comes from (Allen 2018). The left is usually divided into a vast number of groups that can be described in reference to an ideology such as Maoism, Trotskyism, Anarchism, Stalinism, etc. Instead, what might be more significant than such labels based on some kind of origin and lineage, is what they actually try to do and what they achieve. The likely success of an organisation is influenced not only by the clarity of its concepts, but is also strongly affected by context, thus movements must take context into consideration and generate effective practices. Failure on the left is likely, not only because we have the wrong ideas, but because the state is strong. Jean Allen who coined the phrase 'organizational materialism' argues that:

> It becomes too easy to imbue emotional narratives into these events, to call them betrayals, to assume that the leadership is, in their heart of hearts, reformist, or evil, and that had I been in charge, things would have been different. This certainly is true, different tactical decisions will be made by different people, and will lead to different conclusions. [But...] these kinds of analysis often abstract away larger factors. [...] Just as the Turin revolt's failure wasn't due to the writings of Georges Sorel, Occupy Wall Street did not fail from an overabundance of Foucault. (Allen 2018)

A materialist analysis, in this view, starts with this insight that state power makes progress difficult for the left. Given the power of the state and capital, concepts don't automatically move the world. The key question for Allen is how to build capacity to grow the counter power necessary to make radical change.

Base-building can be seen as promoting power, so as to increase capacity, taking up the problem posed by Jean Allen. Power is a concept much argued over in the social sciences but there is no doubt that in most countries a minority have more ability to shape the future than the rest of us. We must raise our capacity to resist. However strong our theory, if our forces are weak, our analysis will be largely irrelevant. As temperatures rise, so will attempts to protect those with wealth and power and to make the rest of us pay for the climate crisis. Climate injustice is already becoming dominant. Unless power is redistributed, the majority of us will have little ability to defend our world and lives. A process of empowerment and organisation is necessary. Without a population with agency and energy, we can be shaped to the needs of an elite. We have seen how the debate on climate change has often been reduced to a dialogue between accelerationists and commodifiers. An exchange between, say, Al Gore and Donald Trump, leaves the rest of us without a voice.

Base-building has the potential to turn climate justice into climate reality. Community organisation involves people, builds (hopefully) social cooperation, and gives communities power. Base-building thus creates alternative institutions that provide alternative sources of power. The material practice of community organisation can, potentially, make more people active, promote collective learning and grow communities of resistance. Whether we are advocating electoral politics or revolutionary change, without a community of support, our strategies will lack the power to succeed. Even if they were to succeed, they would be difficult to sustain without mass and deep popular support.

Base-building not only has the potential to build capacity, it can be used to directly create ecosocialist institutions including community gardens, collective repair stations, and socially organised transport. If Marx and Ostrom give us the theory of the commons economy, base-building acts as ecosocialist practice. In the largely African American city of Jackson, Mississippi, base-building takes an explicitly ecosocialist form. Kali Akuno from Jackson argues that a sixth extinction event is happening and a significant reason for this is loss of habitats. To protect other species, he suggests, we need to reduce human impact by using less energy and land, while engaging in 'major ecological restoration'. He also suggests we need to produce food in a more ecological manner,

> I think permaculture points us in the right direction, as does some degree of small-scale agriculture to at least break the stranglehold the monopolies currently have. I also think we will need to maximize urban density, fairly significantly, to enable more habitats to be recuperated for other species and to restore ecological balance and the replenishment of the soil, which are major carbon sinks. In doing this we will have to turn our urban spaces into 'living farms' to address many of our caloric needs. (Akuno 2019)

Part of the Cooperation Jackson approach has been to fight and win local elections in the city, and aim to implement a local climate action plan. The Cooperation Jackson plan to reach zero emissions includes the following elements: weatherization and energy efficiency retrofitting, solar-thermal energy production, a zero-emissions vehicle fleet, along with expanded and sustainable public transportation (Cooperation Jackson 2015). The Jackson model, as outlined above, blurs the boundary between dual power and conventional politics, with its push to control city government rather than building an entirely autonomous alternative.

Incidentally, a personal project of mine has been to promote climate change adaption and mitigation on my parish council, in Winkfield, Berkshire. This is a large rural and conservative community but I have been keen to build support for serious climate action, involving the local community, other councillors and council staff. We have looked at everything we do and sought both to reduce carbon emissions and to prepare for climate change. Much of this isn't particularly radical but it emphasises that, between personal action and fundamental transformation, collective institutions can achieve something now. For example, an energy survey of all council properties spotted problems with the parish office, as a result more insulation is to be fitted, and we will have a warmer office with lower emissions and lower bills. Local government has its limits, especially in Britain, where services have been cut decade after decades. More radical alternatives are necessary, but we need to take action today. Looking to do what we can with existing institutions can be part of this. However, new institutions are needed, which brings us back to the dual power and base-building strategy.

The base-building strategy is not without its problems, even in a pure and autonomous form. It requires years of patient political organising, but we know that time is short. In turn, many on the left have long argued that you cannot build little islands of utopia in a raging sea of capitalism. Base-building might also act as a way of providing services for communities in the face of government cut backs, bandaging the wounds instead of providing a weapon to challenge the system.

Before the First World War, the established Socialist Parties in Europe, most of whom followed Marx's teachings, built huge organisational capacity. They constructed cooperatives, sports facilities, newspapers, provided welfare for millions and built an alternative. Such was their devotion to the socialist society they were building that they were loath to sacrifice such an alternative; if they had opposed the war, they would have been labelled traitors. As such they risked having their cooperative property taken and their projects dissolved. Community economics in this case created assets that actually tied the socialists to the existing state rather than truly building a community alternative. Of course in Russia the existence of the soviets, community forms of organisation by workers and soldiers, apparently spurred those like Lenin into more radical action (Harding 1980). When the Bolshevik revolution occurred, it had a base. Discussion of the politics of revolution in 1914 and 1917 is a vast and contested topic, I am in no position to explore it here. My point is that it illustrates that dual power models of organisation do not automatically translate into desirable and radical social change. Like all tactics, base-building/dual power is only one way of doing things and should not be transformed into dogma. Nonetheless I think it has some promise; we need to promote base-building as an important part of the practical politics of climate change, in terms of both resistance and adaption.

Revolutionary changes does not look like an immediate option in most parts of the world, so we need to organise in its absence. However, with climate chaos and the crises of capitalism, potential revolutionary situations are likely to become more common. Unless organisational capacity is built, these situations are likely to lead to failure, repression and the strengthening of various forms of elite totalitarianism. One thinks of the example of Egypt, where the Arab Spring displaced a dictator, but after various episodes and events a hard authoritarian system has been restored. Base-building would not have guaranteed victory for the left in Egypt, but it might have made defeat slightly less likely. In Rojava the fact that there was a pre-existing base of Kurdish organisation meant that an uprising was more likely to succeed than in parts of Syria where alternative institutional power was far weaker (Knapp *et al* 2016). Institution building seeds and grows alternative forms of organisation that can potentially create new forms of governance, distinct from the existing state and market hybrid. It also potentially makes communities of resistance more coherent and resilient. In a future likely to be marked by increasing levels of instability, the authoritarian right are organising and preparing, those of us on the left need to do so too.

The base-building concept has some elasticity and might link to a number of projects and approaches that are already evolving. Base trade unions, for example, are being created in many parts of the world, aimed at the self-organisation of workers in sectors of the economy dominated by zero-hour contracts and precarious working conditions. One aspect of this has the been the revival of the International Workers of the World (IWW); known as the Wobblies, that once had hundreds of thousands of members in the USA (Buhle 2005). In Britain, new grassroots unions have been used by otherwise unorganised cleaners, security guards and other poorly paid university workers to successfully gain permanent contracts (Staton 2020). In the USA they have also been active in unionising workers in the fast food industry (Burley 2019). Acorn* is a union for tenants, using mass actions to challenge abusive landlords, unscrupulous letting agents and banks who discriminate against the vulnerable. It has an agenda that goes beyond housing and intersects with wider struggles. For example, its first branch in Bristol campaigned to reverse £8.5 million of planned council cuts to welfare benefits (O'Halloran 2019). None of this necessarily precludes support for more traditional trade unions, or indeed tenants' groups. Many advocates of base building are critical of existing established trade unions, seeing them as corrupt and reformist. While this is not my view, I appreciate that it is held by others. Base-building perspectives are plural and to some extent diverse.

Social centres are another potential element of base-building. Anarchists, animal rights activists and others have a repertoire of squatting empty buildings and turning them into community centres. These provide space for meetings, advice services, canteens and even dance halls. They directly provide capacity for radical organisations, a place for those organising protest to meet and to sleep over before early morning direct action perhaps. An excellent example is the Cowley Club in Brighton, which houses a café, bookshop, and members' bar. It acts as 'a base for those involved in grassroots social change' (Cowley Club 2019a). Established in 2003, volunteers repaired the building

* https://acorntheunion.org.uk/about/

and loans were raised from cooperative finance sources to buy the property. Anna Campbell, who died in Rojava fighting against the Turkish invasion, was a regular user of the Cowley Club. At the time of writing, the Club are planning a mural in her memory. The Club is named after a working-class militant, Harry Cowley. From the 1920s to his death in the 1970s, he helped organise the unemployed, and assisted homeless people to move into squatted buildings after both the first and second world wars. During the 1930s he confronted Brighton's fascists. Combining militancy with practical organising, he seems to have gone on decade after decade with his activism,

> He also campaigned for cheap food, mobilised pensioners, was involved in running social events and social centres and generally organising whatever was needed to provide practical aid for the poor and disadvantaged of the town. His actions were based in local neighbourhoods and outside political parties. We have named the club after Harry to uphold this tradition of grassroots organising and class solidarity. (Cowley Club 2019b)

Brighton has a strong Green Party, as I write, with Britain's only Green Party MP, Caroline Lucas. The local Labour Party have also recently moved to the left, and Brighton is famous as a centre for the LGBTIQ community. The Cowley Club is a small part of a diverse radical community, but by having a physical presence builds continuity and capacity. I suspect that the pre-existence of radical communities has helped the Cowley Club, but the Club also may help reproduce and promote such communities.

Another approach that might fit with base-building is the creation of transition towns (Hopkins 2014). Transition towns create ecological projects and promote practical action on climate change at a grassroots community level. The base-building project from a Marxist or Anarchist perspective is different in an important way from transition towns because it aims to generate resistance. Ultimately, what is being built is a community of activists, whose mutual inks become stronger and whose capacity for challenging existing power structures grows. Existing power structures need to be challenged and this will involve conflict. The notion of transition is also problematic without a close consideration of culture and institutions embedded within an entire society. We should as a thought exercise consider what Lenin might have said about transition towns. If we cannot change the local without changing the whole of society perhaps we will never change anything. Like transition towns, base-building to some extent is local, yet strives to build capacity for a more fundamental challenge to entrenched power and an embedded system.

What is to be done?

My basic case, repeated and repeated once again, is that climate change is a political question. Political in the sense that power relationships drive climate change and shape our responses to it. Abstract good intentions are not enough, political struggle must be engaged in.

Directly challenging corporations who find fossil fuel production profitable is an immediate and vital part of the necessary politics. This slows the destruction and, if successful, removes the pen from those who would write the human story only for short term shareholder gain. From opposing extraction, giving solidarity to Indigenous people, to disrupting the cultural activities and funding of fossil fuel companies, we can directly and immediately intervene to slow climate change.

However, as well as slowing the devastation, and without being proscriptive, we do need to replace capitalism with a new system. Capitalism has proved destructive of life, and as such cannot be defended. This doesn't mean the imposition of our own particular detailed utopia. Ecotopia may be an aspiration but blueprints can be oppressive, there is plenty to do without dreaming them up and trying to paint them onto other peoples' houses. The politics of shoe-horning the rest of society into one's particular plan is innately oppressive. We need instead to grow and sustain a post-capitalist economy based on diversity, participatory politics, democratic ownership, ecological sustainability and shared use. This is technically possible, yet the culture, economics and politics are lagging behind. While one tactic amongst others, the base-building approach can help us to begin to seed and grow a new mode of production. It has several benefits. For example, it is rooted in diversity and experimentation, a flexible approach promoting what works in practice rather than in abstract theory or aggressive polemical dogmas. When successful, it directly increases political capacity, drawing more of us to radical activism and deepening our understanding and commitment. Revolutionary ecosocialism needs to be built by millions of us. Left organising that mobilises tens, hundreds or even thousands is inadequate to the tasks we face. Base-building also works to create community institutions that allow us to better deal with the crisis that is already erupting. We can help and sustain each other. It has disadvantages, and like any broad tactic is not a panacea, but it provides an approach that goes beyond the reforms of electoral politics and the white heat of insurrection. As such I think it is useful for building communities of resistance, engaging in practical work and creating alternatives to established institutions.

Business as usual, the calm water of political consensus and normality, is an extremism that threatens both humanity and the rest of nature. Climate change is to some extent a technical problem because technological innovations mean we can produce and consume energy in a more sustainable way. Yet it is rooted in political considerations and ultimately generated by an entire social system. Félix Guattari, back in 1989, noted:

> So, wherever we turn, there is the same nagging paradox: on the one hand, the continuous development of new techno-scientific means to potentially resolve the dominant ecological issues and reinstate socially useful activities on the surface of the planet, and, on the other hand, the inability of

organised social forces and constitute subjective formations to take hold of these resources to make them work. (Guattari 2014: 20)

This paradox has grown more obvious since Guattari's death in 1992. The practical politics of the climate crisis demands that we organise and transform society. Not an easy task but a necessary one. Like Jean Cavaillès, we must take action, informed by strong concepts, to defend life.

Further Reading

Harding, N. (1980) *Lenin's Political Thought: Theory and Practice in the Democratic and Socialist Revolutions*. 2 vols. London: Macmillan.

A book on Lenin recommended as further reading on climate change might seem odd. Well if like me you believe capitalism is a key driver of climate change, you might want to investigate Marxism in more depth. Harding made me rethink my view on Lenin, this is a very persuasive account of the strengths and weaknesses of Lenin, suggesting his theory is still useful today.

Hulme, M. (2009) *Why We Disagree About Climate Change: Understanding Controversy, Inaction and Opportunity*. Cambridge: Cambridge University Press.

An interesting account of how and why we frame climate change in different ways.

Marshall, G. (2014). *Don't even think about it: Why our brains are wired to ignore climate change*. London: Bloomsbury.

Marshall's book is a veritable encyclopaedia of the science of denial and climate change silence. I would highly recommend reading it, reflecting on its points, re-reading it and looking at the academic texts it cites.

Ravven, H. (2013) *The Self Beyond Itself: An alternative history of ethics, the new brain sciences, and the myth of free will*. New York: The New Press.

I cannot recommend this too highly, a very good account of the ethics of human action, we only have free will to make decisions to the extent that we understand how our beliefs and behaviour are shaped collectively.

Thornett, A. (2019) *Facing the Apocalypse: Arguments for Ecosocialism*. London: Resistance Books.

I don't agree with all of this, in fact it contains a polemical exchange between myself and Alan Thornett on the politics of population. Nonetheless it is an exhaustive and excellent account of the relationship between Marxism and ecology, together with a central focus on ecosocialist action on climate change. It is an engaging book covering many important debates.

Vollmann, W. (2019a) *No Immediate Danger*. London: Penguin.

Vollmann, W. (2019b) *No Good Alternative*. London: Penguin.

William T. Vollmann's two volumes of *Carbon Ideologies*, part travelogue, part chemistry lesson, provides an exhaustive account of the rival effects of oil, coal, natural gas and nuclear power on the environment. At over 1200 pages, this pessimistic masterpiece goes into far more detail in exploring the ideology and science of planetary destruction than anything else I have seen. He describes from the coalface, the fracking rig and the oil field, forms of extraction that are exhausting creation.

Wall, D. (2017) *Elinor Ostrom's Rules for Radicals*. London: Pluto Press.

The economics of ecology, cooperation and trust, an account of how the first woman to win a Nobel in economics worked to nurture ecological and democratic solutions. This teases out a lot of my thoughts on the construction of an economy beyond markets and states building on Elinor Ostrom's impressive work.

Abbreviations

AFD Alliance for Germany, a German right wing political party.
API American Petroleum Institute.
BLF Builders Labourers' Federation, an Australian trade union.
CCS Carbon capture and storage.
CDU Christian Democrat Union, a German centre right political party.
COP Conference of the Parties, the supreme decision-making body of a United Nations convention, here it refers to the climate change convention,
DSA Democratic Socialists of America.
DUP Democratic Unionist Party, a right wing political party in the North of Ireland.
EF! Earth First! a radical environmental network committed to taking direct action.
EGP European Green Party.
GDP Gross Domestic Product.
GND Green New Deal.
HDP Peoples' Democratic Party, a left wing political party in Turkey.
IPCC Intergovernmental Panel on Climate Change.
IWW International Workers of the World.
MEP Member of the European Parliament.
NASA National Aeronautics and Space Administration.
PKK Kurdistan Workers' Party, a revolutionary party.

PRI Partido Revolucionario Institucional, a right wing Mexican political party.

REDD Reducing emissions from deforestation and degradation, it is promoted as part of the United Nations Framework Convention on Climate Change.

RHI Renewable Heat Initiative, a scheme to encourage energy generation from wood pellets in the North of Ireland.

RMT National Union of Rail, Maritime and Transport Workers or Resource Mobilisation Theory.

RTS Reclaim the Streets, an anti-car protest movement.

SPD Social Democratic Party, a German centre left political party.

TD *Teachta Dála*, a member of the Irish Parliament.

TUC Trade Union Congress.

XR Extinction Rebellion.

YPG People's Protection Units, mainly Kurdish Syrian community defence force.

Glossary

Albedo. This refers to the reflectivity of a surface. White surfaces reflect more heat than dark surfaces. As temperatures increase, ice and snow with greater albedo are more likely to melt, replaced by darker surfaces of rock and soil which absorb more heat. This is an example of positive feedback that tends to increase temperatures to a greater extent as a result of climate change.

Base-building. Grassroots organising to build community capacity for resistance.

Commons. Forms of shared collective ownership that include common land, physical goods, information and culture.

Dual Power. The creation of alternative institutions to the state.

Ecology. The science of relationships between organisms and their environments.

Ecosocialism. Ecological socialism

Empirical. Information based on experience and observation.

Epistemology. The study of knowledge which examines how we know what we know.

Malthusianism. The view derived from the economist Thomas Malthus (1766-1834) that over population leads to poverty.

Milankovitch cycles. These are changes in the Earth's orbit that affect climate, generally recognised as a key source of past climate change.

Negative Feedback. Changes that tend to restore a system to its existing state, compensating for external shocks. For example, climate change might lead to warmer temperatures and feedback into less energy use, which reduces carbon emissions. Increased energy use for air conditioning might act as positive feedback increasing emissions.

Object-orientated ontology. A philosophy that focusses on the existence and value of nonhuman objects.

Ontology. The study of being.

Organisational materialism. The principle that the material effects of action are more important than the origin of the ideology motivating actors.

Permafrost. Permanently frozen subsoil found mainly in northern Canada and Russia.

Positive Feedback. Changes that lead to further changes in the same direction, for example, the albedo effect leads to accelerating climate change.

Positivism. An approach that suggests scientific and certain knowledge is based on sensory evidence interpreted via logical analysis.

Bibliography

Abbey, E. (1975) *The Monkey Wrench Gang*. Philadelphia: Lippincott Williams & Wilkins.

Agren, D. (2015) *In Mexican Politics, the Greens Are 'Corruption Turned Into a Party'*. Available at: https://www.vice.com/en_us/article/wja8qb/in-mexican-politics-the-greens-are-corruption-turned-into-a-party (Accessed: 30 March 2020).

Ahmed, N. (2019) *The flawed social science behind Extinction Rebellion's change strategy*. Available at: https://medium.com/insurge-intelligence/the-flawed-science-behind-extinction-rebellions-change-strategy-af077b9abb4d (Accessed: 24 February 2020).

Akuno, K. (2019) *'It's Eco-Socialism or Death', an interview with Kali Akuno*. Available at: https://jacobinmag.com/2019/02/kali-akuno-interview-climate-change-cooperation-jackson (Accessed: 29 March 2020).

Alderman, L. (2019) *What worries Iceland? A world without ice. It is preparing*. Available at: https://www.nytimes.com/2019/08/09/business/iceland-ice-melt-global-warming-climate-change.html (Accessed: 29 March 2020).

Allen, J. (2018) *Organizational Materialism: Considerations on Contemporary Leftism*. Available at: https://theleftwind.wordpress.com/2018/03/30/organizational-materialism-considerations-on-contemporary-leftism/ (Accessed: 29 March 2020).

American Petroleum Institute (2020) *Members*. Available at: https://www.api.org/membership/members#E (Accessed: 30 March 2020).

Amnesty International (2018) *Kenya: Sengwer evictions from Embobut Forest flawed and illegal.* Available at: https://www.amnesty.org/en/latest/news/2018/05/kenya-sengwer-evictions-from-embobut-forest-flawed-and-illegal/ (Accessed: 30 March 2020).

Anon (2019a) *Chinese mines: At least 14 dead in latest disaster.* Available at: https://www.bbc.co.uk/news/world-asia-china-50818647 (Accessed: 30 March 2020).

Anon (2019b) *Music, the Arts and the Climate Protests.* Available at: https://journalofmusic.com/news-uk/music-arts-and-climate-protests (Accessed: 30 March 2020).

Anon (2019c) *'Climate apartheid' between rich and poor looms, UN expert warns.* Available at: https://www.bbc.co.uk/news/world-48755154 (Accessed: 30 March 2020).

Anon (2019d) *Bear falls asleep in wardrobe after entering home.* Available at: www.bbc.co.uk/news/world-us-canada-48731021 (Accessed: 1 April 2020).

Aronoff, K., Battistoni, A., Cohen, D. and Riofrancos, T. (2019) *A Planet to Win: Why We Need a Green New Deal.* London: Verso.

Aubrey, C. (1991) *Meltdown: the Collapse of the Nuclear Dream.* London: Collins and Brown.

Bahro, R. (1986) *Building The Green Movement.* Heretic: London.

Bakalar, N. (2018) *Every 202,500 Years, Earth Wanders in a New Direction.* Available at: https://www.nytimes.com/2018/05/21/science/earth-orbit-change.html (Accessed: 29 March 2020).

Bearzi, G. (2009) 'When Swordfish Conservation Biologists Eat Swordfish', *Conservation Biology*, 23, 1: 1-2.

Behringer, W. (2010) *A Cultural History of Climate.* London: Polity.

Bell, B. (2020) *Austria's Greens get used to a share of power.* Available at: https://www.bbc.co.uk/news/world-europe-51383838 (Accessed: 17 April 2020).

Bengtsson, L. and Hammer, C. (2004) *Geosphere-Biosphere Interactions and Climate.* Cambridge: Cambridge University Press.

Benton, G. (1992) *Mountain Fires: The Red Army's Three-Year War in South China, 1934–1938.* Berkeley: University of California Press.

Barberis, P., McHugh, J. and Tyldesley, M. (2000) *Encyclopedia of British and Irish Political Organizations: Parties, Groups and Movements of the 20th Century.* London: A&C Black.

Berners-Lee, M. (2011) *How Bad are Bananas? The Carbon Footprint of Everything.* Vancouver, Canada: Greystone Books.

Berners-Lee, M. (2019) *What can I do to stop climate change?* Available at: https://newint.org/features/2019/07/01/can-i-do-stop-climate-change (Accessed: 30 March 2020).

Biehl, J. (2015) *Ecology or Catastrophe: The Life of Murray Bookchin.* New York: Oxford University Press.

Blanco, H. (2009) *From the other side of the world Hugo Blanco thanks the Vestas workers.* Available at: https://socialistresistance.org/from-the-other-side-of-the-world-hugo-blanco-thanks-the-vestas-workers/622 (Accessed: 9 April 2020).

Bohrer, A. (2019) *Marxism and Intersectionality: Race, Gender, Class and Sexuality under Contemporary Capitalism.* Bielefeld, Germany: Transcript.

Bookchin, M. and Foreman, D. (1991) *Defending the Earth: A Dialogue Between Murray Bookchin and Dave Foreman.* Boston: South End Press.

Borges, J. (1999) *Collected Fictions.* New York: Penguin.

BP or not BP (2015) *Protesters take to the stage at RSC over BP sponsorship.* Available at: https://bp-or-not-bp.org/2012/04/23/protesters-take-to-the-stage-at-rsc-over-bp-sponsorship/ (Accessed: 30 March 2020).

Breslow, J. (2012) *Robert Brulle: Inside the Climate Change 'Countermovement'.* Available at: https://www.pbs.org/wgbh/frontline/article/robert-brulle-inside-the-climate-change-countermovement/ (Accessed: 30 March 2020).

Brewer, T. (1980) *Marxist Theories of Imperialism: A Critical Survey.* London: Routledge.

Brooker, P. (1988) *Bertolt Brecht: Dialectics, Poetry, Politics.* London: Croom Helm.

Bryson, R. (1976) 'Preface' to Ponte, L. *The Cooling*. Englewood Cliffs, NJ: Prentice-Hall.

Buhle, P. (2005) *Wobblies: A Graphic History of the Industrial Workers of the World*. London: Verso.

Burgin, A. (2009) 'The radical conservatism of Frank H. Knight', *Modern Intellectual History*, 6, 3: 513-538.

Berglund, O. and Schmidt, D. (2020) *Extinction Rebellion and Climate Change Activism*. London: Palgrave Macmillan.

Burgmann, M. and Burgmann, V. (1998) *Green Bans, Red Union: Environmental Activism and the New South Wales Builders Labourers' Federation*. Sydney, Australia: University of New South Wales Press.

Burley, S. (2019) *The Little Big Union joins the growing movement to transform fast food*. Available at: https://wagingnonviolence.org/2019/04/little-big-union-joins-growing-movement-transform-fast-food/ (Accessed: 17 April 2020).

Burns, W. (2013) 'Climate Geoengineering' in Burns and Strauss (eds.) *Climate Change Geoengineering: Philosophical Perspectives, Legal Issues, and Governance Frameworks*. Cambridge: Cambridge University Press.

Burroughs, W. (2003) *Climate: Into the 21st Century*. Cambridge: Cambridge University Press.

Campbell, S., Hunt, C., Scourse, J., Keen, D. and Stephens, N. (2012) *Quaternary of South-West England*. New York: Springer.

Canguilhem, G. (1996) *Vie et Mort de Jean Cavaillès*. Paris: Allia.

Cant, C. (2019) *Deliveroo workers launch new strike wave*. Available at: https://notesfrombelow.org/article/deliveroo-workers-launch-new-strike-wave (Accessed: 30 March 2020).

Chenoweth, E. and Stephan, M. (2011) *Why Civil Disobedience Works: The Strategic Logic of Nonviolent Conflict*. New York: Columbia University Press.

Churchill, W. and Vander Wall, J. (2002) *Agents of Repression: The FBI's Secret Wars Against the Black Panther Party and the American Indian Movement*. Boston: South End Press.

Climate Change Trade Union Group (2014) *One Million Climate Jobs*. Available at: https://www.cacctu.org.uk/sites/data/

files/Docs/one_million_climate_jobs_2014.pdf (Accessed: 30 March 2020).

Cockburn, H. (2019) *Extinction Rebellion founder cleared of vandalism in landmark case after arguing climate change justification.* Available at: https://www.independent.co.uk/environment/extinction-rebellion-court-vandalism-roger-hallam-kings-college-southwark-crown-a8907796.html (Accessed: 30 March 2020).

Cooper, D. (2018) *It's All About That Base: A Dossier on the Base-Building Trend.* Available at: https://theleftwind.wordpress.com/2018/03/16/its-all-about-that-base-a-dossier-on-the-base-building-trend/ (Accessed: 17 April 2020).

Cooperation Jackson (2015) *The Jackson Just Transition Plan.* Available at: https://cooperationjackson.org/blog/2015/11/10/the-jackson-just-transition-plan (Accessed: 30 March 2020).

Cowley Club (2019a) *About the Cowley Club.* Available at: https://cowley.club/about/ (Accessed: 30 March 2020).

Cowley Club (2019b) *Frequently Asked Questions.* Available at: https://cowley.club/about/frequently-asked-questions/ (Accessed: 30 March 2020).

Creighton, J. (2018) *The 'Father of Artificial Intelligence' Says singularity is 30 Years Away.* Available at: https://futurism.com/father-artificial-intelligence-singularity-decades-away (Accessed: 4 April 2020).

Crenshaw, K. (1989) 'Demarginalizing the Intersection of Race and Sex: A Black Feminist Critique of Antidiscrimination Doctrine, Feminist Theory and Antiracist Politics', *University of Chicago Legal Forum*, 140: 139-167.

Cuffe, C. (2007) *Great to Be Back.* Available at: https://cuffestreet.blogspot.com/2007_05_01_archive.html (Accessed: 30 March 2020).

D'Alisa, G. (2019) *Emergenciocracy: why demanding the 'climate emergency' is risky.* Available at: https://undisciplinedenvironments.org/2019/11/21/emergenciocracy-the-risk-of-demanding-a-declaration-of-climate-emergency/ (Accessed: 30 March 2020).

Delbert, C. (2020) *Elon Musk's Battery Farm Is an Undeniable Success.* Available at: https://www.popularmechanics.com/science/a31350880/elon-musk-battery-farm (Accessed: 30 March 2020).

Dickinson, K. (2019) *Are we confusing money with well-being? New Zealand's leaders believe so.* Available at: https://bigthink.com/politics-current-affairs/new-zealand-wellbeing-budget (Accessed: 30 March 2020).

Die Grünen (1983) *Programme of the German Green Party.* London, Heretic.

Doherty, B. (1992) 'The fundi-realo controversy: an analysis of four European Green parties', *Environmental Politics*, 1, 1:95-120.

Doran, P. (2017) *A Political Economy of Attention, Mindfulness and Consumerism.* London: Routledge.

Doyle, K. (1988) *1976: The fight for useful work at Lucas Aerospace.* Available at: https://libcom.org/history/1976-the-fight-for-useful-work-at-lucas-aerospace (Accessed: 30 March 2020).

Duncan, C. (2019) *Revealed: How UK banks are 'threatening' humanity with £25bn funding for dying coal industry.* Available at: https://www.independent.co.uk/environment/climate-change-banks-coal-funding-greenpeace-extinction-rebellion-a9232971.html (Accessed: 29 March 2020).

Dunlap, R. and Jacques, P. (2013). 'Climate change denial books and conservative think tanks: Exploring the connection', *American Behavioral Scientist*, 57, 6: 699–731.

Economist (2018) *The transgender populist fighting fascists with face glitter.* Available at: https://www.economist.com/open-future/2018/12/21/the-transgender-populist-fighting-fascists-with-face-glitter (Accessed: 30 March 2020).

Economist (2019a) *Will Europe's Green parties be the new leaders of the political left.* Available at: https://www.economist.com/europe/2019/01/05/will-europes-green-parties-be-the-new-leaders-of-the-political-left (Accessed: 18 April 2020).

Economist (2019b) *Capuchin monkeys have been using stone tools for around 3,000 years.* Available at: https://www.

economist.com/science-and-technology/2019/06/27/capuchin-monkeys-have-been-using-stone-tools-for-around-3000-years (Accessed: 30 March 2020).

Economist (2020a) *Using satellites to spot industry's methane leaks.* Available at: https://www.economist.com/science-and-technology/2020/02/01/using-satellites-to-spot-industrys-methane-leaks (Accessed: 30 March 2020).

Economist (2020b) *Nuclear power plants are coming to the battlefield.* Available at: https://www.economist.com/science-and-technology/2020/03/12/nuclear-power-plants-are-coming-to-the-battlefield (Accessed: 30 March 2020).

Einwohner, R. and Reger, J. (2008) *Identity Work in Social Movements.* Minneapolis: University of Minnesota Press.

Elbaum, M. (2018) *Revolution in the Air: Sixties Radicals turn to Lenin, Mao and Che.* London: Verso.

Engels, F. (1909) *The Origin of the State, the Family and Private Property.* Chicago: Charles H. Kerr and Co.

Engels, F. (1958) *The Condition of the Working Class in England.* Stanford, CA: Stanford University Press.

Estes, N. (2019) *Our History is the Future: Standing Rock Versus the Dakota Access Pipeline.* London: Verso.

Extinction Rebellion (2020) *Community groups.* Available at: https://rebellion.earth/act-now/resources/communities/community-groups/ (Accessed: 29 March 2020).

Fanon, F. (1983) *The Wretched of the Earth.* Harmondsworth: Penguin.

Fegan, J. (2020) *Election 2020: Who spends what to get your vote?* Available at: www.irishexaminer.com/breakingnews/specialreports/election-2020-who-spends-what-to-get-your-vote-976389.html (Accessed: 9 April 2020).

Ferriéres, G. (2003) *Jean Cavaillès: Un philosophe dans la guerre, 1903-1944.* Paris: Éditions du Félin.

Feyerabend, P. (1975) *Against Method: Outline of an Anarchistic Theory of Knowledge.* London: New Left Books.

Fisher, M. (2013) *Exiting the Vampire Castle.* Available at: https://www.opendemocracy.net/en/opendemocracyuk/exiting-vampire-castle/ (Accessed: 12 April 2020).

Foreman, D. and Haywood, W. (1993) *Ecodefense: A Field Guide to Monkeywrenching*. Chico, CA: Abbzug Press.

Fortuna, C. (2020) *Responding to climate change denial: No, CO2 did NOT rise in the past as it's doing today*. Available at: http://redgreenandblue.org/2020/03/29/responding-climate-change-denial-no-co2-not-rise-past-today/ (Accessed: 12 April 2020).

Foster, J. (2000) *Marx's Ecology: Materialism and Nature*. New York: NYU Press.

Foster, J., York, R. and Clark, B. (2011) *The Ecological Rift: Capitalism's War on the Earth*. New York: NYU Press.

Foyster, G. (2019) *How Abbott still haunts climate policy*. Available at: https://www.eurekastreet.com.au/article/how-abbott-still-haunts-climate-policy (Accessed: 17 April 2020).

France 24 (2019) *Germany's far-right AfD warms to climate change denial*. Available at: https://www.france24.com/en/20190520-germanys-far-right-afd-warms-climate-change-denial (Accessed: 29 March 2020).

Freud, S. (2001) 'Jensen's Gradiva and Other Works (1906 - 1908)' in *Complete Psychological Works of Sigmund Freud. Volume Nine*. New York: Random House.

Freud, S. (2010) *The Interpretation of Dreams*. New York: Basic Books.

Friedman, J. (2019) *Power without Knowledge: A Critique of Technocracy*. Oxford: Oxford University Press.

Gardiner, B. (2019) *For Europe's Far-Right Parties, Climate Is a New Battleground*. Available at: https://e360.yale.edu/features/for-europes-far-right-parties-climate-is-a-new-battleground (Accessed: 1 April 2020).

Goffman, E. (1975) *Frame Analysis: An Essay on the Organization of Experience*. Harmondsworth: Penguin.

Goldsmith, E. (1972) *Blueprint for Survival*. Harmondsworth: Penguin.

Gordon, T. and Todorova, M. (2019) *Future Studies and Counterfactual Analysis: Seeds of the Future*. New York: Springer.

Gorz, A. (1980) *Ecology as Politics*. Montreal, Quebec: Black Rose Books.

Gorz, A. (1982) *Farewell to the Working Class An Essay on Post-Industrial Socialism.* London: Pluto Press.

Granovetter, M. (1973) 'The strength of weak ties', *The American Journal of Sociology,* 78, 6: 1360-1380.

Green and Black Cross (2019) *Statement on Extinction Rebellion.* Available at: https://greenandblackcross.org/statement-on-extinction-rebellion-xr-why-we-can-no-longer-work-with-xr-organiser/ (Accessed: 29 March 2020).

Greenfield, A. (2017) *Radical Technologies. The Design of Everyday Life.* London: Verso.

Griffith, H. (2019) *Why climate 'paralysis' looms over Australia's election.* Available at: https://www.bbc.co.uk/news/world-australia-48145505 (Accessed: 30 March 2020).

Guattari, F. (2014) *The Three Ecologies.* London: Bloomsbury.

Gustin, G. and Henninger, M. (2019) *Central America's choice: Pray for rain or migrate.* Available at: https://www.nbcnews.com/news/latino/central-america-drying-farmers-face-choice-pray-rain-or-leave-n1027346 (Accessed: 29 March 2020).

Hall, J. and Brownstein, B. (2019) *ContraPoints Is the Opposite of the Internet.* Available at: https://www.vice.com/en_us/article/qvygkv/contrapoints-interview-2019-natalie-wynn (Accessed: 30 March 2020).

Hallam, R. (2019) *Common Sense for the 21st Century.* Carmarthenshire, Wales: Turnaround Publishers Services.

Hamaker, J. (1982) *The Survival of Civilization.* Michigan California: Hamaker-Weaver Publishers.

Hamilton, C. (2013) 'What history can teach us about climate denial' in Weintrobe, S. (2013) *Engaging with Climate Change: Psychoanalytical and Interdisciplinary Perspectives.* London: Routledge.

Harding, N. (1980) *Lenin's Political Thought: Theory and Practice in the Democratic and Socialist Revolutions.* 2 vols. London: Macmillan.

Harman, G. (2009) *Prince of Networks: Bruno Latour and metaphysics.* Melbourne: Repress.

Hay, C. (2002) *Political Analysis: A Critical Introduction.* London: Palgrave.

Hayek, F. (1937) 'Economics and Knowledge', *Economica*, 4: 33-54.

Henriksson, L. (2009) *Convert the ailing car industry!* Available at: http://www.internationalviewpoint.org/spip.php?article1761 (Accessed: 30 March 2020).

Hertsgaard, M. (2010) *While Washington Slept*. Available at: https://www.vanityfair.com/news/2006/05/warming200605 (Accessed: 30 March 2020).

Heyden, T. (2019) *How to make biodegradable 'plastic' from cactus juice*. Available at https://www.bbc.co.uk/programmes/p07c2cfz (Accessed: 29 March 2020).

Hochschild, A. (1983) *The Managed Heart: Commercialization of Human Feeling*. Berkeley: University of California Press.

Hoffman, A. (2015) *How Culture Shapes The Climate Change Debate*. Stanford, CA: Stanford University Press.

Holquist, M. (1990) *Dialogism: Bakhtin and His World*. London: Routledge.

Holthaus, E. (2015) *Hot Dam*. Available at: www.slate.com/business/2015/06/the-future-of-hydroelectricity-its-not-good.html (Accessed: 2 April 2020).

Hopkins, R. (2014) *The Transition Handbook: From Oil Dependency to Local Resilience*. Cambridge: Uit Cambridge Limited.

Hudson, M. (2017) *Biographical Availability*. Available at: https://marchudson.net/citizenship/social-movements/biographical-availability/ (Accessed: 30 March 2020).

Hudson, M. (2018) *Excellent Event: Ambiguous transformations: governance, democracy, climate transitions*. Available at: https://marchudson.net/2018/09/20/excellent-event-ambiguous-transformations-governance-democracy-climate-transitions/ (Accessed: 30 March 2020).

Hudson, M. (2019) *The Emotacycle*. Available at: https://marchudson.net/2019/09/23/the-emotacycle-what-it-is-why-it-matters-what-is-to-be-done/ (Accessed: 30 March 2020).

Hulme, M. (2009) *Why We Disagree About Climate Change: Understanding Controversy, Inaction and Opportunity*. Cambridge: Cambridge University Press.

Hulme, M. (2014) *Can Science Fix Climate Change?: A Case Against Climate Engineering.* London: John Wiley.

Hülsberg, W. (1988) *The German Greens: A Social and Political Profile.* London: Verso.

Humphrys, E. (2012) *Neoliberals on bikes.* Available at: https://overland.org.au/2012/07/neoliberals-on-bikes (Accessed: 30 March 2020).

Imbrie, J. (2013) *Ice Ages: Solving the Mystery.* London: Palgrave.

InfluenceMap (2019a) *About InfluenceMap.* Available at: https://influencemap.org/ (Accessed: 30 March 2020).

InfluenceMap (2019b) *Big Oil's Real Agenda on Climate Change.* Available at: https://influencemap.org/report/How-Big-Oil-Continues-to-Oppose-the-Paris-Agreement-38212275958aa21196dae3b76220bddc (Accessed: 17 April 2020).

Inglehart, R. (1977) *The Silent Revolution: Changing Values and Political Styles Among Western Publics.* Princeton, NJ: Princeton University Press.

Jackson, R. (2018) *The Ascent of John Tyndall: Victorian Scientist, Mountaineer, and Public Intellectual.* Oxford: Oxford University Press.

Jackson, T. (2017) *Prosperity without Growth: Foundations for the Economy of Tomorrow.* London: Routledge.

Jameson, F. (1998) *Brecht and Method.* London: Verso.

Jameson, F. (2005) *Archaeologies of the Future: The Desire Called Utopia and Other Science Fictions.* London: Verso.

Jasper, J. (2011) 'Emotions and social movements: Twenty years of theory and research', *Annual Review of Sociology*, 37: 285-303.

Jones, G. (2019) *Viral Socialism: learning from Extinction Rebellion.* Available at: https://www.patreon.com/posts/viral-socialism-26021853

(Accessed: 30 March 2020).

Kahneman, D. (2012) *Thinking Fast and Slow.* London: Penguin.

Keating, D. (2019) *EU decarbonization plan for 2050 collapses after Polish veto.* Available at: www.forbes.com/sites/davekeating/2019/06/20/eu-decarbonisation-plan-for-2050-collapses-after-polish-veto/#68c601bc30b2 (Accessed: 30 March 2020).

Kelly, J. (2018) *Contemporary Trotskyism: Parties, Sects and Social Movements in Britain.* London: Routledge.

Keyes, R. (1985) 'Remineralisation and Soil Fertility', *Green Line*, 30, p.16.

Keynes, J.M. (1936) *The General Theory of Employment, Interest, and Money.* London: Macmillan.

Kingsland, S. (2005) *The Evolution of American Ecology, 1890-2000.* Baltimore, MD: Johns Hopkins University Press.

Kitching, C., Glaze, B., Dourou, S. and Doherty, C. (2019) *Extinction Rebellion activists attacked as commuters drag them off roof of train.* Available at: https://www.mirror.co.uk/news/uk-news/breaking-angry-commuters-drag-extinction-20638767 (Accessed: 2 April 2020).

Knapp, M., Flach, A., Ayboga, E. and Abdullah, A. (2016) *Revolution in Rojava: Democratic Autonomy and Women's Liberation in Syrian Kurdistan.* London: Pluto Press.

Komoto, K. and Kurokawa, K. (2008) *Energy from the Desert: Very Large Scale Photovoltaic Systems: Socio-economic, Financial, Technical and Environmental Aspects.* London: Earthscan.

Kovel, J. (2007) *The Enemy of Nature: The End of Capitalism or the End of the World?* London: Zed.

Kropotkin, P. (1943) *The State: Its Historic Role.* London: Freedom Press.

Kuhn, T. (1962) *The Structure of Scientific Revolutions.* Chicago: University of Chicago Press.

Laclau, E. and Mouffe, C. (1985) *Hegemony and Socialist Strategy.* London: Verso.

Laker-Mansfield, C. (2015) *Is the Green Party the answer?* Available at: https://www.socialistparty.org.uk/articles/20002/28-01-2015/is-the-green-party-the-answer (Accessed: 30 March 2020).

Lanham, R. (2006) *The Economics of Attention: Style and Substance in the Age of Information.* Chicago: University of Chicago Press.

Leach, M. and Scones, I. (2015) *Carbon Conflicts and Forest Landscapes in Africa.* London: Routledge.

Lenin, V. (1961) 'What Is To Be Done?' in *Collected Work*, 5, London: Lawrence and Wishart.

Lenin, V. (1965) *The State and Revolution*. Peking: Foreign Language Press.

Lennon, M. (2019) *No Silver Bullets*. Available at: https://jacobinmag.com/2019/04/green-new-deal-black-radical-tradition (Accessed: 29 March 2020).

Levitz, E. (2016) *The Obama Administration's $1 Billion Giveaway to the Private Prison Industry*. Available at: https://nymag.com/intelligencer/2016/08/obamas-usd1-billion-giveaway-to-the-private-prison-industry.html (Accessed: 15 April 2020).

Lewis, P. and Evans, R. (2013) *Undercover: The true story of Britain's secret police*. London: Faber and Faber.

Liao, S., Sandberg, A. and Roache, R. (2012) 'Human engineering and climate change', *Ethics, Policy & Environment*, 15, 2: 206-21.

Liu, Z. (2015) *Global Energy Connection*. Cambridge, MA: Academic Press.

Lohmann, L. (2006) *Carbon Trading: A Critical Conversation on Climate Change, Privatization and Power*. Uppsala, Sweden: Dag Hammarskjöld Foundation.

Lovelock, J. (2007) *The Revenge of Gaia: Why the Earth is Fighting Back and How We Can Still Save Humanity*. London: Penguin.

Machiavelli, N. (2019) *The Prince*. Cambridge: Cambridge University Press.

Malm, A. (2016) *Fossil Capital: The Rise of Steam Power and the Roots of Global Warming*. London: Verso.

Malm, A. (2018) *The Progress of This Storm: Nature and Society in a Warming World*. London: Verso.

Malsbury, E. (2020) *Samples from famed 19th century voyage reveal 'shocking' effects of ocean acidification*. Available at: https://www.sciencemag.org/news/2020/02/plankton-shells-have-become-dangerously-thin-acidifying-oceans-are-blame (Accessed: 31 March 2020).

Mann, M. and Brockopp, J. (2019) *You can't save the climate by going vegan. Corporate polluters must be held accountable*. Available at: https://eu.usatoday.com/story/opinion/2019/06/03/climate-change-requires-collective-action-more-than-single-acts-column/1275965001/ (Accessed: 30 March 2020).

Mark, C. (2019) *ContraPoints is political philosophy made for YouTube.* Available at: https://www.theatlantic.com/entertainment/archive/2019/01/contrapoints-political-philosophy-natalie-wynn-youtube/579532/ (Accessed: 30 March 2020).

Marshall, A. (2010) *Climate change the 40 year delay between cause and effect.* Available at: https://skepticalscience.com/Climate-Change-The-40-Year-Delay-Between-Cause-and-Effect.html (Accessed: 30 March 2020).

Marshall, G. (2014) *Don't Even Think About It: Why our brains are wired to ignore climate change.* London: Bloomsbury.

Martin, G. (2015) *Understanding Social Movements.* London: Routledge.

Marx, K. (1977) *Capital: A critique of political economy. Volume one.* Harmondsworth: Penguin.

McAdam, D. (1986) 'Recruitment to high-risk activism: The case of Freedom Summer', *American Journal of Sociology*, 92, 1: 64-90.

McBride, S. (2019) *Burned: The Inside Story of the 'Cash-for-Ash' Scandal and Northern Ireland's Secretive New Elite.* Dublin: Irish Academic Press.

McColl, P. (2020) *Free public transport: Why the Scottish Greens' win is the first step to creating a new economy.* Available at: http://bright-green.org/2020/02/28/free-public-transport-why-the-scottish-greens-win-is-the-first-step-to-creating-a-new-economy/ (Accessed: 30 March 2020).

McDowell, I. (2018) *RHI Inquiry: Cash-for-ash - the story so far.* Available at: https://www.bbc.co.uk/news/uk-northern-ireland-45396818 (Accessed: 30 March 2020).

McGrath, M. (2018a) *Climate change: COP24 fails to adopt key scientific report.* Available at: https://www.bbc.co.uk/news/science-environment-46496967 (Accessed: 29 March 2020).

McGrath, M. (2018b) *Climate change: 'Hothouse Earth' risks even if CO2 emissions slashed.* Available at: https://www.bbc.co.uk/news/science-environment-45084144 (Accessed: 29 March 2020).

McIntosh, A. (2008) *Hell and High Water: climate change, hope and the human condition.* Edinburgh: Birlinn.

McLean, I. (2009) 'Lessons from the Aberfan Disaster and its Aftermath', *British Academy Review*, 12: 49–52.

McQueeney, P. and Parkinson, D. (2019) *Building Revolution in the USA: Notes on Marxist Center Conference, 2018.* Available at: https://cosmonaut.blog/2019/01/12/building-revolution-in-the-usa-notes-on-marxist-center-conference-2018/ (Accessed: 30 March 2020).

Melucci, A. (1996) *Challenging Codes: Collective Action in the Information Age.* Cambridge: Cambridge University Press.

Mill, J. (1888) *Principles of Political Economy: With Some of Their Applications to Social Philosophy.* London: Longmans, Green and Co.

Monroe, R. (2020) *Climate destabilization unlikely to cause methane 'burp'* Available at: https://scripps.ucsd.edu/news/climate-destabilization-unlikely-cause-methane-burp (Accessed: 29 March 2020).

Morton, A. (2019) *The climate change election: where do the parties stand on the environment?* Available at: https://www.theguardian.com/australia-news/2019/may/12/the-climate-change-election-where-do-the-parties-stand-on-the-environment (Accessed: 14 April 2020).

Morton, T. (2007) *Ecology without Nature: rethinking environmental aesthetics.* Cambridge, MA: Harvard University Press.

Naess, A. (1993) *Spinoza and the Deep Ecology Movement.* Utrecht, Netherlands: Eburon.

Nagel, S. (2017) *Neoliberals with wind farms. An interview with Marcel Andreu and Gareth Dale.* Available at: https://jacobinmag.com/2017/08/neoliberals-with-wind-farms/ (Accessed: 17 April 2020).

NASA (2019) *Climate Change: How Do We Know?* Available at: https://climate.nasa.gov/evidence (Accessed: 30 March 2020).

Natural History Museum (2019) *Leading scientists set out resource challenge of meeting net zero emissions in the UK by 2050.* Available at: https://www.nhm.ac.uk/press-office/press-releases/leading-scientists-set-out-resource-challenge-of-meeting-net-zer.html (Accessed: 29 March 2020).

Neale, M. (2019) *Leonardo DiCaprio responds to bizarre claim that he bankrolled Amazon fires.* Available at: https://www.nme.com/news/film/leonardo-dicaprio-responds-bizarre-claim-amazon-fires-2582300 (Accessed: 30 March 2020).

Newman, S. (2018) *Political Theology: A critical introduction.* London: Wiley.

Norgaard, K. (2011) *Living in Denial: Climate Change, Emotions, and Everyday Life.* Massachusetts: MIT Press.

Nola, R. and Sankey, H. (2007) *Theories of Scientific Method: an Introduction.* London: Routledge.

O'Halloran, K. (2019) *What Germany's renters union can teach its radical British counterparts.* Available at: https://www.citymetric.com/politics/what-germany-s-renters-union-can-teach-its-radical-british-counterparts-4504 (Accessed: 30 March 2020).

Oikonomakis, L. (2019) *Political Strategies and Social Movements in Latin America: The Zapatistas and Bolivian Cocaleros.* London: Palgrave Macmillan.

Orange, D. (2017) *Climate Crisis, Psychoanalysis, and Radical Ethics.* London: Routledge.

Oreskes, N. and Conway, E. (2012) *Merchants of Doubt: How a Handful of Scientists Obscured the Truth on Issues from Tobacco Smoke to Global Warming.* London: Bloomsbury.

Oreskes, N. and Conway, E. (2014) *The Collapse of Western Civilization: A View from the Future.* New York: Columbia University Press.

Orsman, B. (2020) *Fossil fuels blacklisted by Government for default KiwiSaver funds.* Available at: https://www.nzherald.co.nz/business/news/article.cfm?c_id=3&objectid=12312814 (Accessed: 30 March 2020).

Ostrom, E. (1990) *Governing the Commons: The Evolution of Institutions for Collective Action.* Cambridge: Cambridge University Press.

Ostrom, E. (2012) *Green from the grassroots.* Available at: https://www.project-syndicate.org/commentary/green-from-the-grassroots (Accessed: 24 April 2020).

Paul, S. (2019) *In coal we trust: Australian voters back PM Morrison's faith in fossil fuel.* Available at: https://www.reuters.com/article/us-australia-election-energy/in-coal-we-trust-australias-voters-back-pm-morrisons-faith-in-fossil-fuel-idUSKCN1SP06F (Accessed: 30 March 2020).

Paulson, S. (2018) *The Critical Zone of Science and Politics: An Interview with Bruno Latour.* Available at: https://lareviewofbooks.org/article/the-critical-zone-of-science-and-politics-an-interview-with-bruno-latour/ (Accessed: 30 March 2020).

Pearce, F. (2006) *The Last Generation: How Nature will Take her Revenge for Climate Change.* London: Transworld.

Peden, K. (2014) *Spinoza Contra Phenomenology: French Rationalism from Cavaillès to Deleuze.* Stanford, CA: Stanford University Press.

Philly Socialists (2018) *So, what does it mean for us to be a 'base-building' socialist organization?* Available at: https://threadreaderapp.com/thread/1042468757074046978.html (Accessed: 30 March 2020).

Pindar, I. and Sutton, P. (2014) 'Translators Introduction' in Guattari, F. *The Three Ecologies.* London: Bloomsbury.

Piper, J. (2019) *Oil and gas lobby split by Trump administration rollback of federal methane rules.* Available at: https://www.opensecrets.org/news/2019/08/oil-gas-lobby-split-by-trump-rollback-of-methane-rules/ (Accessed: 30 March 2020).

Ponte, L. (1976) *The Cooling: Has the Next Ice Age Already Begun?* Englewood Cliffs, NJ: Prentice-Hall.

Powell, J. (2011) *The Inquisition of Climate Science.* New York: Columbia University Press.

Proust, M. (2000) *In Search of Lost Time: III The Guermantes Way.* Vintage Books: London.

Pryde, P. (1991) *Environmental Management in the Soviet Union*. Cambridge: Cambridge University Press.

Punk Academic (2019) *Extinction Rebellion: Credit, Criticism & Cops*. Available at: http://criticallegalthinking.com/2019/04/29/extinction-rebellion-credit-criticism-cops/ (Accessed: 30 March 2020).

Rahm, D. (2014) *Climate Change Policy in the United States: The Science, the Politics and the Prospects for Change*. Jefferson, NC: McFarland.

Rapp, D. (2012) *Ice Ages and Interglacials: Measurements, Interpretation, and Models*. New York: Springer.

Ravven, H. (2013) *The Self Beyond Itself: An alternative history of ethics, the new brain sciences, and the myth of free will*. New York: The New Press.

Reay, D., Smith, P. and van Amstel, A. (2010) *Methane and Climate Change*. London: Earthscan.

Research & Degrowth (2019) *Definition*. Available at: https://degrowth.org/definition-2/ (Accessed: 30 March 2020).

Richard, W. (2018) *UK air pollution could cause 36,000 deaths a year*. Available at: https://www.kcl.ac.uk/news/uk-air-pollution-could-cause-36000-deaths-a-year (Accessed: 30 March 2020).

Ritholtz, B. (2019) *Labor Market Is Doing Fine With Higher Minimum Wages*. Available at: www.bloomberg.com/opinion/articles/2019-01-24/u-s-economy-higher-minimum-wages-haven-t-increased-unemployment (Accessed: 5 April 2020).

Robert, A. (2018) *COP24: Return of king coal*. Available at: https://www.euractiv.com/section/climate-environment/news/cop24-return-of-king-coal/1299369/ (Accessed: 30 March 2020).

Rostain, S. (2012) *Islands in the Rainforest: Landscape Management in Pre-Columbian Amazonia*. Walnut Creek, CA: Left Coast Press.

Ruddiman, W. (2003) 'The Anthropogenic Greenhouse Era Began Thousands of Years Ago', *Climatic Change*, 61: 261–293.

Ryan, B. (2018) *Lung Cancer Health Disparities*. Oxford: Oxford University Press.

Ryan, G. (2017) *The troubling 'tactics' politicians are using to attack rooftop solar*. Available at: https://www.cnbc.com/2017/09/18/attacking-rooftop-solar-energy-is-a-big-mistake-commentary.html (Accessed: 29 March 2020).
Saltmarsh, C. (2018) *5 Reasons I'm Not Joining the 'Extinction Rebellion'* Available at: https://novaramedia.com/2018/11/18/5-reasons-im-not-joining-the-extinction-rebellion/ (Accessed: 29 March 2020).
Santer, B. (1996) *Letters to the Editor: No Deception in Global Warming Report*. Available at: https://www.wsj.com/articles/SB835642517444561000 (Accessed: 30 March 2020).
Santer, B. (2019) *My climate story: Ben Santer*. Available at: https://www.climateone.org/audio/my-climate-story-ben-santer (Accessed: 2 April 2020).
Scott, J. (1998) *Seeing like a State: How Certain Schemes to Improve the Human Condition Have Failed*. New Haven: Yale University Press.
Seitz, F. (1996) *A major deception on 'global warming'*. Available at: https://www.wsj.com/articles/SB834512411338954000 (Accessed: 30 March 2020).
Simms, A., Pettifor, A., Lucas, C., Secrett, C., Hines, C., Leggett, J., Elliot, L., Murphy, R., and Juniper, T. (2008) *A Green New Deal: Joined-up policies to solve the triple crunch of the credit crisis, climate change and high oil prices*. London: New Economics Foundation.
Sinfield, A. (1998) *Gay and After: Gender, Culture and Consumption*. London: Serpent's Tail.
Skeptical Science (2011) *What is the net feedback from clouds?* Available at: https://www.skepticalscience.com/clouds-negative-feedback.htm (Accessed: 30 March 2020).
Smith, G. (2009) *Emissions Trading? I beg your indulgence*. Available at: https://www.earthisland.org/journal/index.php/magazine/entry/emissions_trading_i_beg_your_indulgence/ (Accessed: 30 March 2020).
Smith, R. (2016) *Gaston Bachelard: Philosopher of Science and Imagination*. New York: State University of New York Press.

Snow, D., Rochford, B., Worden, S. and Benford, R. (1986) 'Frame Alignment Processes, Micromobilization, and Movement Participation', *American Sociological Review*, 51: 464-481.

Sommerlad, J. (2018) *California wildfires: Why Trump's claim 'forest management' is to blame is completely wrong - and why he'll keep saying it*. Available at: https://www.independent.co.uk/environment/california-wildfires-trump-tweet-malibu-camp-fire-woolsey-hill-ventura-county-a8629801.html (Accessed: 29 March 2020).

SourceWatch (2009) *The Frank Statement*. Available at: https://www.sourcewatch.org/index.php?title=The_Frank_Statement (Accessed: 15 April 2020).

Spinoza, B. (1930) *Ethics: Demonstrated in geometrical order and divided into five parts*. Oxford: Oxford University Press.

Sprague de Camp, L. (1968) *The Great Monkey Trial*. New York: Doubleday.

Spratt, D. and Dunlop, I. (2019) *Existential climate-related security risk: a scenario approach*. Melbourne, Australia: National Centre for Climate Restoration. Available at: https://docs.wixstatic.com/ugd/148cb0_90dc2a2637f348edae45943a88da04d4.pdf (Accessed: 31 March 2020).

Staton, B. (2020) *The upstart unions taking on the gig economy and outsourcing*. Available at: https://www.ft.com/content/576c68ea-3784-11ea-a6d3-9a26f8c3cba4 (Accessed: 17 April 2020).

Stauffer, R. (1957) 'Haeckel, Darwin, and Ecology', *The Quarterly Review of Biology*, 32, 2: 138-144.

Stephenson, M. (2013) *Returning Carbon to Nature: Coal, Carbon Capture, and Storage*. Amsterdam, Netherlands: Elsevier.

Sturrock, J. (1979) *Structuralism and Since: From Lévi Strauss to Derrida*. Oxford: Oxford University Press.

Sun, S. (2019) *Stop Using Phony Science to Justify Transphobia: Actual research shows that sex is anything but binary*. Available at: https://blogs.scientificamerican.com/voices/stop-using-phony-science-to-justify-transphobia/ (Accessed: 29 March 2020).

Surin, K. (1990) 'Marxism(s) and "The Withering Away of the State"', *Social Text*, 27: 35–54.

Sweeney, S. (2013) *Resist, Reclaim, Restructure: Unions and the Struggle for Energy Democracy*. New York: Rosa Luxemburg Stiftung.

Symons, J. (2019a) *The DNC's climate problems run deep*. Available at: https://thehill.com/opinion/energy-environment/449177-the-dncs-climate-problems-run-deep (Accessed: 31 March 2020).

Symons, J. (2019b) *Ecomodernism: Technology, Politics and The Climate Crisis*. London: John Wiley.

Tarrow, S. (1998) *Power in Movement*. New York: Cambridge University Press.

Taylor, D. (2014) *Toxic Communities: Environmental Racism, Industrial Pollution, and Residential Mobility*. New York: NYU Press.

Taylor, M. (2018) *15 environmental protesters arrested at civil disobedience campaign in London*. Available at: https://www.theguardian.com/environment/2018/oct/31/15-environmental-protesters-arrested-at-civil-disobedience-campaign-in-london (Accessed: 2 April 2020).

Thacker, P. (2006) 'The Many Travails of Ben Santer', *Environmental Science & Technology*, 40, 19: 5834-5837.

Thompson, L. (2010) 'Climate change: the evidence and our options', *The Behavior Analyst*, 33, 1153-170.

Thornett, A. (2019) *Facing the Apocalypse: Arguments for Ecosocialism*. London: Resistance Books.

Trade Union Congress (2010) *Congress Decisions 2010*. Available at: https://www.tuc.org.uk/research-analysis/reports/congress-decisions-2010 (Accessed: 2 April 2020).

Trinder, M. (2020) *Cuba found to be the most sustainably developed country in the world*. Available at: www.greenleft.org.au/content/cuba-found-be-most-sustainably-developed-country-world (Accessed: 10 April 2020).

Vaughan, A. (2019) *Bitcoin's climate change impact may be much smaller than we thought*. Available at: https://www.newscientist.

com/article/2224037-bitcoins-climate-change-impact-may-be-much-smaller-than-we-thought/ (Accessed: 29 March 2020).

Vollmann, W. (2019a) *No Immediate Danger*. London: Penguin.

Vollmann, W. (2019b) *No Good Alternative*. London: Penguin.

Walker, R. and Jeraj, S. (2016) *The Rent Trap: How We Fell Into It and How We Get Out of It*. London: Pluto Press.

Wall, D. (1999) *Earth First! and the Anti-Roads Movement*. London: Routledge.

Wall, D. (2008) *Prosperity without Growth: Economics beyond Capitalism*. London: Sustainable Development Commission.

Wall, D. (2009) *Colombia: 'Green fuel' kills*. Available at: www.greenleft.org.au/content/colombia-green-fuel-kills (Accessed: 29 March 2020).

Wall, D. (2017) *Elinor Ostrom's Rules for Radicals*. London: Pluto Press.

Wall, D. (2018) *Hugo Blanco: A Revolutionary for Life*. London: Merlin Press.

Wark, M. (2015) *Molecular Red: Theory for Anthropocene*. London: Verso.

Wells, P. and Touboulic, A. (2017) *Rich and famous lifestyles are damaging the environment in untold ways*. Available at: https://theconversation.com/rich-and-famous-lifestyles-are-damaging-the-environment-in-untold-ways-71641 (Accessed: 29 March 2020).

Wilkinson, T. (2013) *The Rise and Fall of Ancient Egypt*. London: Bloomsbury.

World Bank (2015) *Electricity production from hydroelectric sources (% of total)*. Available at: www.data.worldbank.org/indicator/eg.elc.hyro.zs World Bank 2015al (Accessed: 30 March 2020).

Wynn, N. (2018) *The Apocalypse*. Available at: https://www.youtube.com/watch?v=S6GodWn4XMM (Accessed: 30 March 2020).

Yusoff, K. (2018) *A Billion Black Anthropocenes Or None*. Minneapolis: University of Minnesota Press.

INDEX

Abbott, Tony 101, 121-2
Aberfan, Wales 68
acidification 12, 26
Acorn union 169
Adani coal mine 101
Africa 53, 142
African Americans 66, 120, 145, 161
African National Congress 53
Agenda 21 134
Agriculture 22, 25, 31, 39, 42, 60, 143, 166
air pollution 10, 31
Akuno, Kali 166
albedo 21, 117, 178, 179
Alberta 130
Allen, Jean 164,165
Alliance for Germany 102, 176
alternative communities 94
alternative economic indicators 47
Amazon rainforest 9, 30, 104, 155
American Enterprise Institute 105
American Petroleum Institute 106, 107
Amsterdam 61
anarchism 5, 56-8, 93-4, 98, 110, 139, 147-9, 164, 169-70
Anderson, Kevin 37
Angry Jack effect 133
animal liberation 65, 91, 93, 169
Anthrobscene 159
Anthropocene 159

Antarctic 30
anti-war 36, 91, 93, 94, 97
Aotearoa, New Zealand 47, 53, 54, 61, 161
Arab Spring 148, 168
Aral Sea 147
archaeology 19
Arizona 24
Arctic 27, 30
Argentine 34
Aron, Raymond 7
arrests 77, 78, 85
artificial intelligence 46, 157
Assad, Bashar 148
Aspen, Colorado 15
Atlantic 26, 27
Atlantic City 100
atmosphere 11, 21, 24-8, 42, 111-12, 139
Attenborough, David 33
attention, economics of 34, 36, 58, 77, 85, 86, 87, 90, 93, 121, 122, 125-31, 140, 144, 153
Australia 31, 32, 53, 69, 82, 94, 100, 101, 103, 118, 121, 122, 130, 152, 159, 161
austerity 59, 62
Austria 58, 59, 124, 140, 143
Austrian economics 140, 145
Austrian Green Party 58, 59
Austrian People's Party 59

INDEX

Bachelard, Gaston 109, 110
bacteria 11
badgers 97
Bahro, Rudolf 55
Bakhtin, Mikhail 130
Balibar, Étienne 4
bananas 33-5, 37, 49
Bangkok 30
banks, banking 14, 44, 59, 98, 99, 169
Bangladesh 30, 42
Barrier Reef 101
base-building 87, 161-72, 178
batteries 40, 41
Battlo, Jean 64
bear 156
Bearzi, Giovanni 37
beef 155
beer 38
Belgium 13
Benton, Gregor 91, 92
Berkshire 1, 71, 166, 167
Berners-Lee, Mike 33-5, 37, 49
biofuels 42, 107, 108
biographical availability 85-7, 90
biology 9, 10
biomass boilers 143
biomedical modification 38-9
bitcoin 34
Black Panther Party 161, 162
Blade, Zoë 136
Blair, Tony 57
Blanco, Hugo 4, 64
Bloch, Ernst 157
Blitz, London 125
Blueprint for Survival 54
Bogdanov, Alexander 1
Bolsheviks 140, 147, 168
Bolsonaro, Jair 2, 3, 16, 103, 104, 117, 118, 155
Bookchin, Murray 56, 58, 91, 148, 149

'boom and bust' economic cycle 44
Borges, Jorge Luis 34
Borneo 13
Brazil 2, 3, 16, 21, 53, 103, 104, 117, 118, 155
Brecht, Bertolt 79-80
Brentford, London 73
Brexit 57, 104, 105, 137
Brexit Party 104
Brighton 169, 170
Bright Green 62
Bristol 169
Britain 1, 9, 26, 29, 31, 33, 37, 41, 44, 51, 52, 55, 57, 69, 70, 72, 75-8, 83, 86, 90, 91, 93, 104, 105, 106, 118, 144, 146, 153, 155, 161, 167, 169, 170
British Empire 13, 14
British Museum 155
British Petroleum 107, 153-5
British Psychoanalytical Society 125
Brittany 7
Brockopp, Jonathan 35
Broecker, Wally 28
Brown, Derren 34
Buchanan, James 140
Builders Labourers' Federation 69
Bulgaria 60
buses 51, 61, 102
Butler Creek, Missouri 156

cactus juice 38
Cadogan Hall 155
California 82, 103, 115, 142
Campbell, Anna 170
camps, protest 86, 94
camps, prison 121
Canada 11, 52, 130, 135, 154
Canberra 32
cancer 106, 114, 115, 116
Cancun 60
Canguilhem, Georges 8

Canning Town 79
Cantor, Georg 5
cap and trade 142
capitalism 1, 2, 12, 14, 15, 17, 34, 39, 44-8, 65, 67, 72, 89, 98, 99, 150, 157, 160, 167-71
Capitalocene 159
capuchin monkeys 157
carbon capture 42, 43
carbon dioxide 1, 11-15, 22-6, 31-7, 43, 44, 46, 107, 110-12, 114, 117
carbon footprint 33-7
carbon lock-in 36
carbon sinks 26, 27
carbon tax 44, 121
carbon trading 44, 139, 142, 166
cars 15, 35, 40, 41, 47, 66, 67, 70, 97
cash for ash 143
Catholic Church 111, 131, 142
Cavaillès, Jean 5-8, 18, 110, 173
central planning 140
Cesar Andreu Iglesias Community Garden 164
Challenger, Professor 16
Chiapas 149
China 30, 37, 69, 90, 91, 161
Chennai 30
Chevron 107
Chicago school 140
Chippenham, Wiltshire 1
Christian Democratic Party 52, 104
chlorofluorocarbons 116
city states 149
civil rights 80, 144
class 9, 14, 57, 63, 65-73, 92, 102, 140, 163, 170
climate apartheid 120
climate emergency 29-32, 75, 97, 98
climate justice 43, 62, 78, 165
Clinton, Hilary 161, 162
clouds 26

coal 1, 2, 3, 11, 13, 14, 28, 31, 40, 42, 43, 64, 66, 68-70, 73, 95, 99, 101-3, 106, 108, 119, 120, 143, 155
cobalt 41
coffee 21
cognitive dissonance 125, 126, 152
Cohen, Stanley 129
Cohn-Bendit, Danny 56
Cold War 21, 119, 134
Colombia 42, 53
Colorado 15
comedy 42, 123
commodity 48, 139
commons 4, 14, 47, 51, 61-3, 70, 139, 149, 166, 178
communication 37, 88-9, 121-38
communism 7, 56, 90, 91, 140, 161, 162, 164
Communist Caucus 164
concepts 6, 9, 17, 18, 53, 82, 159, 160, 164, 165, 173
confirmation bias 128
Congo 13
conservatism 9, 54, 57-9, 83, 91, 102, 104, 116, 126, 166
conveyor belt radicalism 152, 153
Cooperation Jackson 164, 166
COP (Conference of the Parties) 3, 5, 102
Copernican revolution 109
copper 41
Corbyn, Jeremy 62
corruption 60, 113, 117, 141-4, 169
Corsham, Wiltshire 1
counter movements 95, 97, 99
Cowley Club, Brighton 169, 170
Cowley, Harry 170
Cowley, Oxfordshire 67
cremation 34
Crenshaw, Kimberlé 66
Crichton, Michael 129
Cronus 136

Chthulucene 159
Cuba 148, 161
Cuffe, Ciarán 58-9
culture 8, 9, 12, 39, 46, 49, 63, 81-4, 95, 97, 99, 109, 124, 128-30, 133, 150, 154, 163, 171, 172
cycles, social movement 82, 86, 87, 119
cycling 2, 38, 47, 58, 62, 72, 73
Czech Republic 101

Dakota 41
D'Alisa, Giacomo 147
dams 41, 44, 151
Daoists 91
dark kitchens 73
Dean, Jodi 15
death penalty 60
death squads 103
decentralisation 53, 149, 150
deep ecology 8, 89, 90, 155
Deepwater Horizon 154-5
DeGraffenreid versus General Motors 66
degrowth 12-14, 34, 39, 46-9, 54, 59, 63, 65
deindustrialisation 72
Deliveroo 73
democracy 43, 53, 148, 149, 157
Democratic Confederalism 148, 149
Democratic Party 45, 106, 128, 161, 162
Democratic Socialists of America 161, 164
Democratic Unionist Party 143
denial 17, 19, 25, 28, 50, 94, 101-22, 123-6, 129, 135, 147, 152, 153, 174
depression, economic 44
deserts 41
desire 16, 47, 92, 93, 133, 146, 147
Desmog UK 104
Devensian 29

DiCaprio, Leonardo 104
diesel 41
Dijon 157
direct action 1, 56, 63, 75-80, 84, 91, 93, 94, 153, 169
Ditfurth, Jutta 56
divestment 61, 77, 93, 154
dodo 31
Doyle, Sir Arthur Conan 16
droughts 15, 40, 82, 95, 100, 128
Dual Power 162-8, 178
Dublin 59
Duda, Andrzej 102
Durham 20, 70
Dutch Green Left (Groen Links) 56, 61
dysprosium 41

Earth First! 76, 83, 91, 93, 94, 96
earthquakes 76
economics 4, 12, 14, 47, 48, 140, 148, 167, 172, 175
economic growth 12-14, 34, 39, 46-9, 54, 59, 63, 65
Ecologist 127
ecology 8-12, 16, 17, 22, 36, 46, 48, 49, 53, 55-8, 64-6, 89, 91, 100, 127, 147-8, 155-8, 166, 172, 173, 175
Ecology Party 22, 53-5, 57, 65
ecological modernism 10, 46
ecosocialism 4, 13, 56, 57,67, 153, 162, 165, 166, 172, 175
Ecuador 13
Edinburgh 61, 155
Einstein, Albert 109, 111, 130
Elbaum, Max 161
elections 52, 56, 58, 59, 61, 63, 76, 83, 104, 106, 108, 166
electricity 36, 40-2, 61, 103
electric cars 41
Embobut Forest 142
emergenciocracy 147

eminent domain 41
emotion 9, 36, 82, 86, 87, 89, 124, 128, 135-7, 164
emotacycle 86
Engels, Friedrich 10, 67-9, 140, 157
environmental justice 165
Environmental Liberation Front 129
environmental racism 120
environmental prisoners 155
Environment Protection Agency 116
epistemological break 110
epistemology 109, 110, 145, 178
Estes, Nick 41, 42
Estonia 101
ethnicity 15, 66, 87, 88, 91
European Elections 52, 59, 104
European Green Party 51-3, 59-61
European Parliament 56, 62, 104, 158
European Union 42, 57, 58, 102, 104, 142
Evangelical People's Party 56
extinction 29, 31, 75, 94, 105, 166
Extinction Rebellion 5, 17, 29, 33, 63, 75-99, 105, 121, 125, 129, 144, 153, 154,160
ExxonMobil 105-8

Facebook 90, 93, 108, 131, 158
Fanon, Frantz 13
fake news 118, 131
feminism 67, 149
Fermanagh 143
fertilizers 42
Feyerabend, Paul 110
Fianna Fáil 58, 158
Fine Gael 58, 158
fingerprints 111, 112
Finland 52
fires 32, 82, 95, 100, 101, 103, 128
First World War 160
flying 36, 37, 59, 124

flying pickets 73
Foote, Eunice 11
Foster, Arlene 143
Foster, John Bellamy 40, 157, 158
Foucault, Michel 92, 164
four-day working week 72
fracking 60, 76, 94, 98, 155, 175
frames, framing 78, 88-90, 91, 95, 98, 118, 121, 122, 127, 129, 130, 132, 134, 139, 152, 174
France 5-7, 16, 52, 55, 92, 109, 132, 146, 149
Franklin, Benjamin 24
Freedom Summer 80, 85
free public transport 51, 61
Freud, Anna 125
Freud, Sigmund 48, 117, 124, 125, 133
Fridays for Future 74, 75, 76, 81, 84, 92
Friedman, Jeffrey 132
friend-enemy distinction 132-5
functionalism 80
fundis 55-6

Gaia hypothesis 23
Galileo 111
gardens 38, 91, 164, 166
gas 11, 13, 24, 28, 31, 42, 43, 69, 105, 106, 154, 155, 175
Gauland, Alexander 102
Gaullism 7
gender 9, 14, 15, 67, 121, 159
General Election (Australia 2019) 100, 101
General Election (Irish 2007) 58
General Election (Irish 2011) 59
General Election (UK 2015) 62
General Election (UK 2019) 57
genetics 38, 39, 46, 115
geoengineering 21, 39, 40, 147
geology 10, 14, 23, 29, 41, 159

INDEX

George C. Marshall Institute 116
German Green Party 52, 53, 56, 58, 59
Germany 7, 10, 52, 53, 56, 58, 59, 79, 102, 124, 130
gig economy 72, 72
Glantz, Stanton 115
Global Climate Coalition 113
global cooling 19-23
global dimming 21, 23
Global Greens 59, 60
Goffman, Erving 88, 89
Goldsmith, Teddy 57
golf 15
Gore, Al 37, 123, 165
Gorz, André 64, 65, 73
Granovetter, Mark 90
grassroots democracy 53, 149, 150
Greater London Assembly 56
Greece 29, 128, 136
Green and Black Cross 78
green bans 69
Greenham Common 93, 94
green jobs 45, 64, 66, 71, 97, 103
Greenland 30
Green Left (Turkey) 53
Green Line 22
Green New Deal 33, 44-6, 49, 62, 71, 147
Green Party 16, 51-63, 64, 83, 104, 144, 158, 170
Green Party of England and Wales 51-63, 64, 83, 104, 170
Greenpeace 76
green strikes 66, 69
greenwash 107
Gross Domestic Product 46
Guanglong mine 69
Guangzhou 30
Guatemala 31
Guattari, Félix 16, 100, 172, 173
Guizhou 69

Gulf Stream 26
Haeckel, Ernst 10
Hallam, Roger 5, 83, 92, 95, 96, 154
Halsema, Femke 61
Hamaker, John D. 22, 23
Hamilton, Clive 130
Haraway, Donna 159
Harman, Graham 156, 157
Hayek, Friedrich von 140
heat pumps 61, 71
Heathrow Airport 52, 153
hedges 9
Hegel, Georg Wilhelm 49, 67, 160
Henriksson, Lars 70
herbicides 60
herbs 38
high modernism 145-7
Hitler, Adolf 125
Hobson, John 13
Ho Chi Minh City 30
Hochschild, Arlie 135
Hoffman, Andrew J. 124, 130, 131, 135, 138
Holmes, Sherlock 16
home brewing 38
homophobia 9, 103, 161
Honduras 31
Hong Kong 30
Hope, Matt 104
Hot house Earth 27, 97
Hudson, Marc 82, 86
Hulme, Mike 20. 174
Hungary 101, 102
hunger strike 93, 154
Hunter's Hill, New South Wales 69
Hurricane Katrina 116
hydrology 10
hydroelectricity 40-2
hypocrisy 37

ice age 11, 19-22, 25
ice cores 28, 29

Iceland 61
identity construction 36-7, 65, 70, 82, 87, 88, 91, 117, 131, 133, 135, 152
ideology 7, 46, 53, 57, 146, 164, 175
imperialism 13, 14, 67, 98
incinerators 120
India 53, 69, 119
Indiana 135
indigenous 4, 13, 41, 42, 104, 135, 142, 145, 149, 155, 171
information deficit model 124, 128
infrastructure 36, 44
Inglehart, Ronald 65
Institutional Revolutionary Party 60
insulation 38, 45, 167
Institute for Public Policy Research 127
interglacials 24, 29
Intergovernmental Panel on Climate Change 106, 113
International Coordinator, Green Party 51, 52
International Workers of the World 169
intersectionality 66, 67, 85
Iran 95
Ireland 26, 36, 52, 54, 57-9, 62, 142-4
Irish Civil War 58
Irish Green Party 54, 58, 59, 158
Irish Republicanism 58, 143
Islamic State, so-called 148, 149
Islamophobia 59, 118, 119
Isle of Wight 64
Italy 52

Jackson, Mississippi 164, 166
Jackson, Tim 34, 46, 47
Jakarta 30
Jakobsdóttir, Katrín 61
Jameson, Frederic 48

Japan 53
Jarvis, Chris 62
Jevons paradox 40
Jewish people 7, 124, 130
Jones, Graham 77
Jupiter 24
justice 17, 43, 53, 62, 67, 78, 165
just transition 70, 71, 73

Kahneman, Daniel 126-7
Katowice 102
Kenya 142
Keynes, John Maynard 44, 45
Keystone 64
King's College 93, 154
Kinnock, Neil 57
KiwiSaver 61
Knight, Frank 140
knowledge problem 140
Koala Hospital 32
Kuhn, Thomas 109
Kurds 41, 53, 90, 148, 149, 160, 168
Kuwait 3

Labor Party 100, 101
Labour Party 57, 61, 62, 83, 161, 170
Labuan, Borneo 13
Laclau, Ernesto 65
Lagos 30
Laki volcano 24
Lakota 41
lamb, roast 122
lasers 36
Latour, Bruno 109, 111, 156-8
Lawrence Livermore National Laboratory 112
Learjet 15, 37
Lenin, Vladimir 1, 13, 140, 146-9, 158-61, 168, 171, 174
Lennon, Myles 45
Lewis, William 4
LGBTIQ+ 170

liberalism 10
Liberal-National Coalition 100
liberation theology 149
lifestyle change 15, 33-8, 46, 97, 133, 134
Limbaugh, Rush 137
lithium carbonate 41
Little Mermaid 136
local climate change plan 71, 72, 166, 167
logic 6, 7, 21, 48, 89, 110, 119, 124, 137, 141
London 7, 35, 51, 52, 56, 60, 77-9, 93, 125, 154
Lorient 7
Lovecraft, H. P. 159
Lovelock, James 23
Lower Brule Sioux Tribe 41
Lucas Plan 71
Luxemburg, Rosa 13

Machiavelli, Niccolò 151
Malcolm X Center 164
Malm, Andreas 13, 14, 120, 156-8, 160
Malthus, Thomas 21, 91, 178
ManBearPig 123
Manchester 68
Manila 30
Mann, Michael 35
manthropocene 159
Maoism 55, 56, 90, 164
Maori 47
Marble Arch 77
Marcuse, Herbert 157
Marshall, George 37, 43, 124-38, 174
Martínez, Jorge 60
Marx, Karl 48, 49, 67, 140, 157, 166
Marxist Center 162-4
Massey Corporation 68
mathematics 5-8
McAdam, Doug 85

McColl, Peter 61, 62
McGuinness, Martin 143
Melbourne 36
methane 11, 25-7, 38, 68, 107
methane hydrates 27
MethaneSAT 38
Mexico 31, 38, 53, 59, 60, 118, 121, 149, 160
Mexican Green Party 59, 60
Michigan 124
Midlands 29
migration 31, 59, 118, 120, 121, 123, 137
Milankovitch cycles 24, 111, 178
Milanković, Milutin 24
military 30, 36, 96, 103
Mill, John Stuart 10
minimum wage 141
Mississippi 166
mode of production 47-50, 73, 97, 98, 138, 150, 172
Modi, Narendra 119
momentum training 77
Montcoal, West Virginia 68
Montag, Warren 4
Moore, Jason 156, 159
Moral hazard 143
Morrison, Scott 66, 101, 117, 121
Morton, Tim 9, 10
Moscow 36
Moses 48
motorways 35, 76, 93, 94, 99
Mouffe, Chantal 65
mountain top removal 69
Movement for Survival 54
Mumbai 30
Mundy, Jack 69
music 81, 82, 136
Mutual Aid 163

Naess, Arne 8
NASA 28

National Academy of Sciences 115
National Coal Board 68
National Gallery 155
Natural History Museum 41
Natural Parks 9, 24
nature 2, 6, 7-10, 12, 14, 16, 19, 27-9, 39, 49, 66, 108, 145, 152, 157, 158, 172
Nazism 5-7, 124, 130
negative feedback 23, 26, 27, 30, 117, 178
neodymium 41
neoliberalism 58, 72, 140
Netherlands 7, 51, 52, 56, 61, 107
Nevada 103
new communist movement 161
New England 149
New Mexico 41
New Orleans 128
new social movements 82, 84
New South Wales 31, 32, 69
Newton, Thomas 109, 111
New York 11, 56, 66, 100, 148,
Nietzsche, Friedrich 134
Nirvana fallacy 43
Noel Coward Theatre 155
nonlinear dynamics 27
nonviolence 53, 56, 63, 75, 91, 93, 95, 96, 144
Nordic Green Left 62
Norgaard, Kari Marie 135
Northern Ireland Executive 142
North Sea 153
Norway 8, 52, 135
Nozick, Robert 53
nuclear power 12, 29, 31, 55, 175
nuclear war 29, 34, 55, 116

Oak Ridge National Laboratory 115
Obama, Barack 121
Object-orientated ontology 156, 157, 179

Öcalan, Abdullah 149
Ocasio-Cortez, Alexandria 45
Occupy 91, 164
oil 1, 11, 13, 20, 28, 31, 40-5, 54, 69, 105-8, 113, 116, 120, 130, 143, 145, 148, 153-5, 175
One million climate jobs 71
Orange, Donna 124
Oregon petition 116
Organization for a Free Society 164
organisational materialism 164, 179
Ostrom, Elinor 4, 11, 47, 70, 149, 150, 166
overdetermination 12, 24, 48
Overton window 62
Oxford 67
Oxford Circus 77
Oxford Street 81
ozone layer 39, 116

Pacific 47
Pacifist Socialist Party 56
palaeoclimatology 25, 29
Palaeolithic 29
Palestine 95
palm oil 42
Pantglas 68
paramilitaries, right wing 42
Parikka, Jussi 159
Parliament Square 77
Paris 7, 103, 104, 107
Paris agreement 103, 104, 107
Parsons, Talcott 80
Party for Socialism and Liberation 164
passenger pigeon 31
Pearce, Fred 28
Peden, Knox 6, 7
pensions 61, 72, 76, 154
People and Planet 78
People Party 54
Peoples' Democratic Party 53

People's Protection Units 149
permaculture 148, 166
permafrost 11, 25, 30, 31, 179
Peru 13, 64, 135
pesticides 42
Petrified Forest 24
petrol 41, 93
Pet Shop Boy 88
Philadelphia 106, 163
Philippines 5, 53, 95
Philly Socialists 163
physics 109, 111, 116, 130
philosophy of science 6-8, 108-111
Piccadilly Circus 77
plastic 12, 31, 38
platform capitalism 72, 73, 90, 136, 145
pleasure 81, 129, 133
podcasts 131, 133, 136
Poland 3
police 78, 80, 84, 87, 96, 144, 145
political opportunity structures 83, 84, 96, 98
Pompeii 68
Ponte, Lowell 20-23, 26
Populism, Right-wing 66, 102, 105, 118, 120
Port Macquarie, New South Wales, Australia 32
Portugal 51
Pope 131, 142
Popper, Karl 110
positive feedback 23, 25-7, 117
positivism 110, 179
post-materialism 65, 82
power, relations 67, 78, 171
pressure groups 76
profit 36, 45-8, 171
property rights 4, 12, 14, 47, 51, 61-3, 70, 139, 149, 166, 178
prophets 10, 48, 124, 125
proportional representation 54, 56

prosperity (without growth) 34, 46, 47, 63
protectionism 103
Proust, Marcel 132
psychoanalysis 16, 48, 117, 124, 125, 133
Ptolemy 109, 111
Puerto Rico 164

REDD 142
RMT trade union 64
racism 14, 17, 67, 78, 80, 118, 120, 121, 145
Radical Party 56
radioactive waste 12, 31, 69
rainforests 9, 30, 39, 104, 155
rational choice 81, 124, 153
rationalism 6-8
Ravven, Heidi 4, 137, 174
Rawls, John 53
Raworth, Kate 167
Raymond, Lee R. 105, 106
Read, Jason 4
Reader's Digest 114
realos 55
Reclaim Shakespeare Company 154, 155
Reclaim the Streets 93
recycling 61
Refoundation Caucus 164
refugees 102, 118, 119
renewable energy 40-2, 45, 55, 70, 71, 101-3, 106, 107, 119, 142, 143, 145, 148
rent controls 141
repair 47, 166
repertoires of contention 84
replication, movement 77
Republican Party 102, 103, 105, 106, 117, 128, 152, 161, 162
Resistance, French 6-8

resource mobilisation theory 80, 81, 83, 87
retrofitting 71, 166
rich, the 15
right-wing ecological politics 57
R.J. Reynolds Tobacco 115
road-rail buses 71
Rockefeller University 115
Rojava 148, 149, 160, 168, 170
Roma people 146
Roosevelt, Franklin 44
Roundhouse 155
Royal Dutch Shell 107
Royal Opera House 155
Royal Shakespeare Company 154, 155
rubber 13
Ruddiman, William 25
Russia 130, 140, 147, 159, 168
Russian Revolution 140, 147, 159, 168

sabotage 93
Saltmarsh, Chris 78
Sami 135
Sanders, Bernie 45, 161
San Francisco 15, 91
Sano, Yeb 5
Santer, Benjamin 111-13, 116
satellite 38, 112
Schmitt, Carl 132, 133
Schumpeter, Joseph 140
Science Museum 155
Scopes Monkey trial 111
Scotland 51
Scottish Green Party 51, 54, 57, 61
Scottish National Party 61
Scottish parliament 56, 61
Scott, James C. 145-7
Seattle 135
Second World War 5-8, 44, 65
Seitz, Frederick 113, 115, 116, 119

Sen, Amartya 53
Sengwer 142
set theory 5
sewage 61
Shakespeare, William 154, 155
Shanghai 30
Shorten, Bill 101
Siberia 11
Sikhs 87
silence, social 128, 129
Simpsons, The 118
Singer, Fred 116, 119
Sinn Féin 62, 143
Social Democratic Party 52, 55
social feedback 117-21
socialism 10, 56, 64, 140, 161-4, 167
Socialist Party USA 164
social justice 53, 62
social media 34, 37, 50, 85, 90, 106, 117, 118, 131, 136
social movements 5, 16, 54, 65, 72, 74-99, 139, 153, 160
social murder 68, 69
soils 11, 22, 26, 27, 166, 178, 179
SolarCity 103
solar power 38, 41, 71, 92, 103, 166
solar system 23, 109, 111
Sorbonne 6
Sorel, George 164
South Africa 53
South Park 123
Soviets 147, 167, 168
Soviet Union 21, 119, 147, 148, 167
soya 155
Spinoza, Baruch 7, 8, 10, 49, 130
spy cops 96, 144
SS Cabinet Minister 42
Stolze, Ted 4
Stalin, Joseph 147
Stalinism 147, 164
Strasbourg 6
Stratford-upon-Avon 154

stratosphere 39
strawberries 38
strong programme 110
structuralism 35
subjectivity 67
sulphate 39
sulphur dioxide 24
sun 23, 24, 40
Sunrun 103
Sussex 16, 169
Sweden 5, 13, 70, 75
Switzerland 52
swordfish 37
Syria 118, 148, 149, 160, 168, 170

Taiwan 53
Tarrow, Sidney 83, 96
Tate Britain 155
Tea Party 133, 134
technological singularity 157
Tempest, The 154
Tennant, Neil 88
Texas 128
text messages 34
Thames 77
theory 4, 5, 17, 35, 109, 158-61, 165, 166, 172, 174
Thompson, Lonnie 29
Thornett, Alan 67, 175
Thunberg, Greta 5, 75, 102, 129
Tianjin 30
Tillerson, Rex 106
tipping points 27, 30, 49
thinking, fast and slow 89, 16
Tobacco 106, 114-16, 119
Tobacco Industry Research Committee 114
Tory-Lib Dem 62
Total 107
trade unions 57, 64-74, 139, 154, 162, 169
Trades Union Congress 71
trains 37, 70, 79, 84, 125

transgender 9
transition towns 170, 171
troposphere 112
Trotsky 147
Trotskyism 90, 164
trout 34
Trump, Donald 2, 3, 4, 16, 45, 59, 66, 100-104, 106, 107, 117, 118, 121, 131, 137, 161, 162, 165
Turkey 41, 53, 148, 149
Twitter 93
Tyndall Centre for Climate Change Research 20
Tyndall, John 11

UKIP Party 104
uncertainty, scientific 118, 127, 130, 131
unemployment 52, 70, 107, 141
unintended consequences 3, 98, 141, 142, 144
universities 6, 20, 34, 41, 61, 115, 124, 153, 154, 169
Upper Big Branch disaster 68
Utopia (and ecotopia) 2, 147, 167, 171
Utrecht 51

values, party 54
veganism 3, 35, 87, 133
Venezuela 40
Venus 24, 101
Vestas 64
Verfremdungseffekt 79
Vermont 161
Victoria Station, London 51
Vienna 124
Viet Cong 149
Vietnam 30, 141, 169
Vivint 103
Vollmann, William T. 42, 68, 73, 175

Wales 29, 41, 68
Wall Street Journal 113
Warringah 101
Wark, McKenzie 1, 159
Washington, D.C. 112
Waterloo Bridge 77
water vapour 26, 111
Weimar Republic 130
West Antarctic Ice Sheet 30
Whisky Galore! 42
Wiltshire 1
wind energy 40, 58, 64, 70, 102

Winkfield, Berkshire 71, 166-7
Winnicott, Donald 125
working class 64-74, 140, 163
Wynn, Natalie (Contrapoints) 123, 131, 133, 136-7

Yellowstone 9
Yusoff, Kathryn 14

Zapatistas 149, 160

Also from Merlin Press

Hugo Blanco
A Revolutionary for Life

Derek Wall

Hugo Blanco is Peru's best-known revolutionary.
A leader of the indigenous people of the Andes, he was born in 1934 in Cusco, the former Inca capital. He is a lifelong environmental campaigner in defence of the natural riches of the Andean region and beyond.
In the 1960s he led a successful armed peasant uprising demanding land rights. He was placed on death row and released only after a huge international campaign supported by Jean-Paul Sartre. In exile in Chile he was lucky to escape death after the 1973 coup.
More recently Hugo Blanco was a Presidential candidate and was elected as a Senator in Peru. He was exiled to Mexico, where he was influenced by the Zapatistas. Still politically active today, he publishes the newspaper Lucha Indigena (Indigenous Struggle).

This engaging political biography surveys the life of this unassuming but compelling activist – a guerrilla fighter praised by Che Guevara, one-time member of the Fourth International – from the 1960s to the present. It is a story of ideas and activism: surveying Hugo Blanco's views on defence of the environment, social and political movements, indigenous peoples, left govern-ments and political strategy.

With 10 black & white photos.

ISBN. 978-0-85036-748-5 paperback

www.merlinpress.co.uk

We The Indians
The indigenous peoples of Peru and the struggle for land

Hugo Blanco

With a foreword by Eduardo Galeano

Eduardo Galeano writes:
'These pages, written in bursts, disorderly, jubilant and desperate, tell of the adventures and misfortunes of the man who headed the campesino struggle in Peru, the organiser of the rural trade unions, the man who pushed for an agrarian reform born from below and fought for from below.'

Hugo Blanco has walked his country forwards and backwards, from the snow-covered mountains to the dry coast, through the rainforests where the tribes are hunted like beasts. And wherever he went, on the way he helped the fallen to get up, and the silent ones to speak.
The authorities accused him of being a terrorist. They were right. He sowed terror among the owners of land and of people. He slept under the stars and in cells occupied by rats. He went on fourteen hunger strikes. In one of them, when he could barely go on any longer, the Minister of the Interior made a kind gesture and sent him a coffin as a gift. More than once, the district attorney demanded the death penalty, and more than once the news was published that Hugo had died.
He continues to be that smart, crazy man who decided to be an Indian, even though he was not, and turned out to be the most Indian of all.
Hugo Blanco was a key protagonist in the events he describes. His vivid and direct language takes the reader on an inspirational journey to the heart of Peru – looking for a respectful relationship with Pacamama (Mother Earth), and with its indigenous communities and their struggles for land reform and change in the 1950s and 1960s.

With 15 woodcuts

ISBN. 978-0-85036-738-6 paperback